Georgij N. Akoev   Yurij N. Andrianov

Sensory Hair Cells

Synaptic Transmission

With 92 Figures

Springer-Verlag
Berlin Heidelberg New York
London Paris Tokyo
Hong Kong Barcelona
Budapest

Dr. Georgij Nikolaevich Akoev
Dr. Yurij Nikolaevich Andrianov
Pavlov Institute of Physiology
Russian Academy of Sciences
Nab. Makarova 6
199034 St. Petersburg
Russia

Title of the original edition: "Synapticheskaja peredacha v rezeptorach akustikolateraljnoj sistemy" "Nauka" 1989.

ISBN-13:978-3-642-77921-3 e-ISBN-13:978-3-642-77919-0
DOI: 10.1007/978-3-642-77919-0

Library of Congress Cataloging-in-Publication Data. Akoev, G.N. (Georgij Nikolaevich) [Sinapticheskaja peredacha v rezeptorach akustikolateraljnoj sistemy. English] Sensory hair cells: synaptic transmission / G.N. Akoev, G.N. Andrianov. p. cm. Includes bibliographical references and index.
ISBN-13:978-3-642-77921-3
1. Labyrinth (Ear) 2. Neural transmission. 3. Afferent pathways. I. Andrianov, G.N. (Yurij Nikolaevich) II. Title.
QP471.A5613 1993 612.8'58-dc20 93-7183 CIP

Softcover reprint of the hardcover 1st edition 1993

Production Editor: Herta Böning, Heidelberg
Typesetting: Best-set, Hong Kong
31/3145-5 4 3 2 1 0 – Printed on acid-free paper

Preface

This monograph summarizes the results of the authors' own investigations as well as relevant literature data about the mechanism of synaptic transmission in sensory hair cells of the acousticolateral system. It deals mainly with primary processes in hair cells. Additionally, data on the properties and selectivity of various ionic channels are presented, with data concerning the mechanism of release of the neurotransmitter from the hair cell and the properties of postsynaptic responses of afferent fibres. The role of excitatory amino acids in synaptic transmission is analyzed.

This book gives a comprehensive overview of current research on synaptic transmission in sensory hair cells, which should be valuable to both those in the field and neuroscientists of various disciplines.

St. Petersburg, January 1993

G. N. Akoev
G. N. Andrianov

Table of Contents

Introduction

The acousticolateral system plays an important role in the biological activity of vertebrates. It includes the organ of Corti, vestibular and otolith apparatus, and lateral-line organs (Ariëns Kappers et al. 1936; Schmahlhausen 1938; Disler 1960; Dijkgraaf 1963; Vinnikov 1971, 1979). The lateral-line organs are located in the epithelium and subcutaneous structures, but they are similar to the receptors of the inner ear. The vertebrate inner ear organ is generally agreed to be an evolutionary end-product of the inward migration and adaptation of the lateral-line system (Baird 1974; Dallos 1984). The auditory and vestibular organs, which together comprise the inner ear, are absent in invertebrates, but the earliest vertebrates such as lamprey already possess a developed vestibular labyrinth. The commom embryonal development and innervation, and the similarity in functional and morphological organization of their peripheral and central parts make it possible to consider the organs of the inner ear and the lateral line as a single acousticolateral system. The great similarity in anatomy and mechanisms for the reception of mechanical stimuli means that data obtained in the study of more accessible sensory receptors of the lateral-line organs can be applied for a better understanding of the receptor mechanisms of the inner ear, the study of which is more difficult in technical terms.

The specialized receptors for the perception of weak electric fields in the lateral-line organs of some lower aquatic vertebrates may also be considered to be a part of the acousticolateral system (Lissmann 1958; Lissmann and Machin 1958). This system has attracted the close attention of investigators ever since its discovery (see reviews by Bennett 1967, 1971; Akoev et al. 1971; Andrianov et al. 1979; Viancour 1979; Bullock 1982; Andrianov and Ilyinsky 1983; Brown and Ilyinsky 1984; Electroreception 1986; Andres and von During 1988). The sources of signals for electroreceptors may be various electric phenomena occurring in the aquatic environment as a result of different physical and physicochemical processes (such as changes in salinity and temperature, water movement in the Earth's magnetic field, etc.), as well as those evoked by numerous biological processes connected with the life of the animals. Such a wide range of factors which may be sources of signals for the electroreceptor system testifies to the important role of this system in animals life.

The attention paid to the problem of electroreception may also be accounted for by the fact that electroreceptors, as derivatives of the lateral-line organs, are relatively simply organized and easily accessible receptors. This, in conjunction with the advantage of using electric current as an adequate stimulus, makes electroreceptors a convenient model for investigating receptor mechanisms. When studying different aspects of electroreception, elasmobranchs (sharks and skates) are widely used. They possess a well-developed and highly sensitive electroreceptor apparatus, the ampullae of Lorenzini.

Receptors of the acousticolateral system belong to secondary sense cells in which the initial transduction is carried out by neuroepithelial cells that transmit synaptically to the afferent nerve fibers. Understanding the mechanisms of synaptic transmission is important from many points of view. It is at the synaptic level that the transformation of local receptor potentials into the impulse activity of afferent nerve fibers takes place. Synaptic transmission provides convergence of a great number of receptor cells onto one afferent fiber and enhancement of receptor sensitivity to adequate stimulation. An important feature of the receptor synapses is their high sensitivity to potential. The mechanisms of the sensory synapse have not been sufficiently studied, particularly the nature of the neurotransmitter released by the receptor cells. Analysis of the mechanisms of synaptic transmission in the receptors of the acousticolateral system may also be relevant to other secondary receptors as well as to the understanding of the main synaptic mechanisms of action in the peripheral and central nervous systems.

In spite of the great interest in the problem of neurotransmitter identification in the receptors of the acousticolateral system (see reviews by Guth et al. 1976; Klinke 1981, 1986; Guth and Melamed 1982; Bobbin et al. 1984; Akoev and Andrianov 1988, 1989), this problem cannot be considered as solved. The data of different authors are often contradictory. To a certain extent, this can be due to differences in the properties of the objects studied and also to the complexity of the methods used for studying the receptors of the inner ear. Of considerable advantage are the electroreceptors (ampullae of Lorenzini) of some fishes, whose large size makes it possible to perform experiments on an isolated preparation.

The first chapter of the present book is devoted to the description of morphological data characterizing the structure of the receptors of the acousticolateral system. In the second chapter the general characteristics of the electrical activity of different receptors are discussed. The third chapter is devoted to the mechanisms of acidic amino acid-using synapses. The fourth chapter presents data of the neuropharmacological analysis of synaptic transmission in the receptors of the acousticolateral system.

Chapter 1

Anatomy of the Mechano- and Electroreceptors of the Acousticolateral System

Throughout the vertebrate acousticolateral system the sensory receptors are hair cells (Fig. 1). They are grouped in specialized sensory regions in the organ of Corti (the organ of hearing), the cristae ampullaris of the semicircular canals (the vestibular apparatus), the macula utriculi, macula sacculi and macula lagenae (the otolith organs), neuromasts and canals (the lateral-line organs); and epidermal cavities (electroreceptors). According to the modern concept, the otolith organs, semicircular canals, and the organ of Corti are united into a labyrinth. Utriculus, sacculus, lagena, and semicircular canals with their ampullae form the vestibular part of the labyrinth as opposed to the cochlea with the organ of Corti, which is related to the acoustic function of the labyrinth. The otolith apparatus provides information about linear inertial forces acting on the head, such as gravity and the forces arising during linear acceleration. The ampullae of the semicircular canals respond to angular accelerations occurring during head rotations. The vestibular system plays an important role in the maintenance and regulation of complicated coordinated reflexes in animals and man. It participates in realizing proprioceptive functions, the coordination of the eyes, head, and body during motor activity. At the same time, vestibular activity is integrated with other sensory systems. The lateral-line organs of cyclostomes, fish, and amphibians may detect local water movements and surface waves. Electroreceptors comprise a morphologically and physiologically heterogeneous group of lateral-line receptors specialized for the detection of electric fields in the aquatic medium. They are found in cyclostomes, most non-teleosts, and some teleosts, in a number of amphibians and in the only known mammal – the Australian platypus (Bullock 1982; Bullock et al. 1983; Andrianov and Ilyinsky 1983; Electroreception 1986; Scheich et al. 1986).

1.1 The Organ of Corti

As shown in numerous studies, the receptor apparatus of the organ of Corti is represented by inner and outer hair cells (Fig. 2). They are distinguished by their location within the organ of Corti, by their relation to the supporting

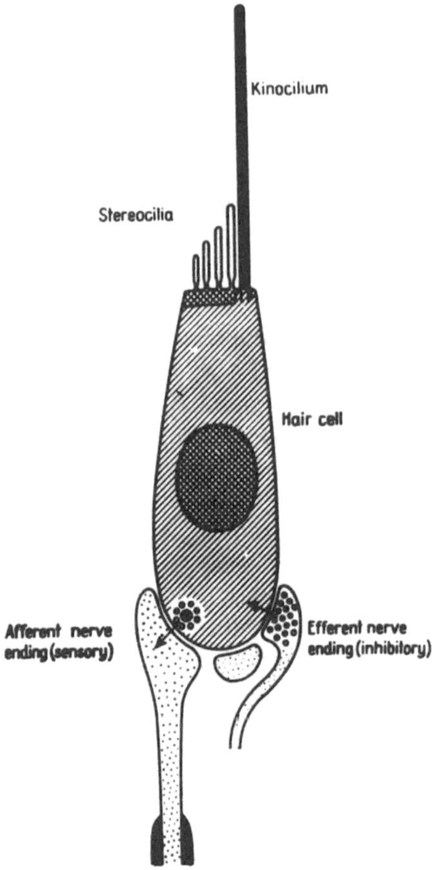

Fig. 1. Schematic diagram showing basic morphological components of the common hair cell. At the apex of the cell body several sterocilia emerge from a cuticular plate, and a single kinocilium is anchored to the intracellular basal body. Presynaptic specializations such as the synaptic bar (*dark circle*) and vesicles (*small open circles*) are seen adjacent to the afferent nerve terminal, while efferent endings themselves are full of vesicles (*small circles*). (Flock 1971)

cells, by the types of innervation, by the relationship of the ciliary apparatus to the tectorial membrane, and by the structure of the cytoplasm. The inner hair cells are distributed on one side of the cochlea tunnel, whereas the outer ones are located on the other side. Hair cells lie on supporting cells whose projections enclose receptor cells. Above the hair cells there is a tectorial membrane which is a gelatinous plate covering the organ of Corti.

A number of authors (Engström and Sjöstrand 1954; Smith and Dempsey 1957; Iurato 1961, etc.; see review by Lim 1986) have shown that both the outer and inner cells possess a cuticular plate at their apical surface, 2 μm thick. A bundle of stereocilia protrude from the cuticular plate. Stereocilia

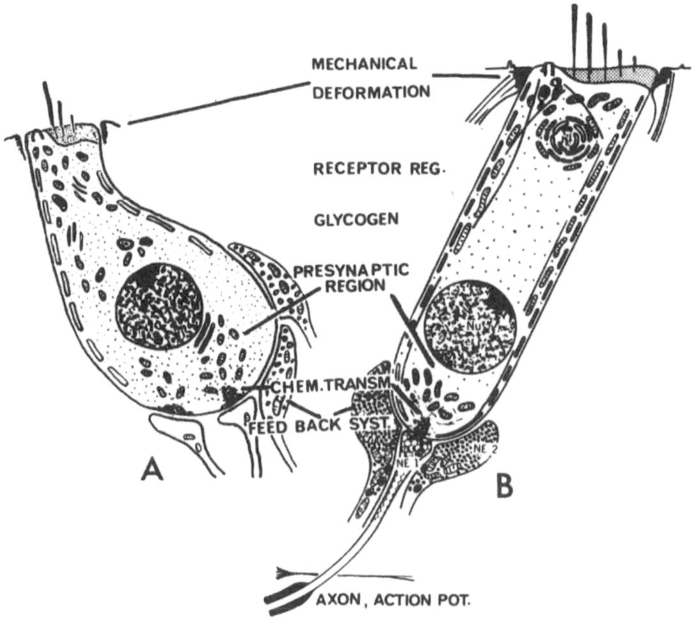

Fig. 2A,B. Schematic drawings of: **A** inner and **B** outer hair cells showing the general principles of their structural organization. (Ades and Engström 1974)

end in rootlets penetrating deep into the cuticle. The cuticle covers most of the distal part of the cell, but leaves a small area on one side open. At this point the cytoplasm of the cell is in direct contact with the plasma membrane. It is oriented radially towards the periphery of the cochlear spiral and contains the basal body. The development of the basal body and the kinocilium connected with it takes place during the embryonic period of life (Bredberg et al. 1972; Russell et al. 1986a). After birth, the kinocilium generally disappears in mammals, only the basal body remaining. Sometimes a rudimentary kinocilium may also remain. For example, on the outer hair cells of some monkeys, several rudimentary kinocilia may be observed throughout the lifetime of the animal (Ades and Engström 1974). In birds, the hair cells of the organ of Corti have a "normal" kinocilium (Vinnikov 1979).

The inner hair cells are one layer thick. Their apical surfaces are provided with a bundle of about 60 stereociclia 0.3 μm in diameter. They consist of several rows, perpendicular to the direction of the cochea radius. The length of stereocilia in the bundle is graded, those closer to the cuticle-free zone being longer (6–7 μm). Their length also increases towards the apex of the spiral.

The outer hair cells are supplied with a hundle of 100–120 stereocilia. They are disposed in a regular three-layer W-form, the base of the W turning towards the periphery of the organ of Corti. This pattern of stereocilia distribution seems to increase the sensitivity of the receptors to certain directions of stimulation. The basal body is always located in the indenture of the W. The stereocilia of the outer hair cells are thinner than those of the inner hair cells (0.15 μm). Their length in the bundle is also different; the length of stereocilia increases towards the periphery of the bundle and also from the base of the cochlea towards its apex.

Stereocilia consist of the fibrillar core surrounded by a triple-layered plasma membrane. A net of thin fibrils links stereocilia with each other (Pickles et al. 1984). The filaments of stereocilia fuse close to the plasma membrane of the receptor cell and form a basal rootlet which protrudes into the cuticle. The rootlet is considerably thinner than the hair proper. The central portion of the hair is a continuation of the rootlet. The peripheral tops of the longest stereocilia of the outer hair cells are in close contact with the tectorial membrane (Kimura 1966). The shorter hairs from the outer hair cells have no contact with the tectorial membrane. The stereocilia of the inner hair cells do not seem to possess such a contact with the tectorial membrane (Lim 1972).

The cytoplasm of hair cells in the organ of Corti is rich in cell organelles including the nucleus, endoplasmatic reticulum, ribosomes, lysosomes, the Golgi apparatus, and numerous mitochondria. These tend to concentrate in the supranuclear zone of the cell towards the cuticle-free region. In the outer hair cell there is strict laminar organization of the endoplasmic reticulum, forming cell cisternae inside the plasma membrane. The presence of these organelles, together with an accumulation of a great number of mitochondria, suggests an active ion exchange in the outer hair cells in the organ of Corti (Dallos 1984). In the basal part of the cells synaptic bodies can be seen surrounded with numerous vesicles. In the outer hair cells synaptic bodies may sometimes be absent.

It is generally agreed that the nerve fibers within the organ of Corti are divided into two classes: afferents and efferents. Afferents are dendrites from spiral ganglion neurons; efferents refer to the axonal endings of neurons located in the brain stem which carry information from the brain to the cochlea. However, this classification needs to be refined in the light of recent data on cochlear anatomy (Pujol and Lenoir 1986).

The inner and the outer hair cells are in contact with a great number of nerve endings (Smith and Sjöstrand 1961). Among them there are small, light bouton-shaped nerve endings scantily supplied with mitochondria and synaptic vesicles, and more numerous and larger dark nerve endings with a great number of synaptic vesicles and mitochondria. At present, at least three types of nerve endings in the organ of Corti are distinguished. The type I nerve endings do not degenerate after transsection of the efferent axon and seem to be afferent. The nerve endings of this type are observed at the bases of both inner and outer hair cells. The afferent dendrite endings on the inner hair cells are usually thick and filled with large mitochondria. In the presynaptic zone of inner hair cells there are ribbon-like synaptic bodies with a cluster of synaptic vesicles which may be up to 800 Å in diameter. Each inner hair cell contacts about 20 afferent nerve endings. The presynaptic membrane at these sites possesses invaginations.

It is curious that only 5% of afferent fibers pass through the tunnel of Corti to the outer hair cells, which constitute more than 75% of the total number of receptor cells in the organ of Corti (Spoendlin 1985). At the basal end of each of the outer hair cells, several nerve endings form synaptic contacts. In outer hair cells, synaptic bodies are not always seen. Instead, the presynaptic region is filled with numerous mitochondria and vesicles. The presence of synaptic vesicles and the broad synaptic cleft (150 Å) are fairly convincing arguments that the afferent synapses are chemical in nature (Pappas 1975; Dallos 1984). In most outer hair cells there are reciprocal synapses when a single nerve ending of a nonvesicular type has two types of synaptic specialization (Nadol 1981). There is no evidence of direct functional connections between afferent fibers innervating the outer and the inner hair cells. In no site of the afferent system of two types of receptor cells are there synaptic contacts even between their axons. Thus, the inner and the outer hair cells of the organ of Corti are distinguished not only quantitatively and structurally but also in their patterns of innervation. For example, the inner hair cells are known to be innervated by the fibers of type I (bipolar) of the spiral ganglion, whereas the outer hair cells are supplied by the processes of type II neurons (pseudounipolar) (Spoendlin 1969; Kiang et al. 1982; Schwartz 1986). The existence of such considerable differences in function and structure supports the probable attribution of the inner and the outer cells to two different sensory systems in the organ of Corti (Spoendlin 1985; Pujol and Lenoir 1986).

Type II nerve endings mainly form axosomatic synapses on the outer hair cells. There is evidence that this type of nerve ending degenerates after selective transection of the efferent olivocochlear bundle (Iurato 1962; Kimura and Wersall 1962; Spoendlin and Gasek 1963). The efferent innervation of the outer hair cells is extremely well developed. Thus, there are from six to ten nerve endings of this kind on each outer hair cell and their total number in all the outer hair cells amounts to 40 000 (Spoendlin 1966). The total area of efferent axo-somatic contacts is approximately ten times greater than that of afferent terminals. Efferent nerve terminals are characterized by a relatively large diameter (0.7–3 μm in guinea pig and up to 3.5 μm in chinchilla) (Engström 1960; Smith and Rasmussen 1965), by numerous synaptic vesicles, 280–375 Å in diameter (Smith 1967), and by the presence of well-developed subsynaptic cisterns along the postsynaptic membrane in the outer hair cell.

Type III vesicular nerve endings form axodendrite synapses in the inner hair cells. These endings also degenerate after a selective transection of the olivocochlear bundle (Spoendlin 1966). Together with type II nerve endings, they constitute an efferent system of hair cells and are connected with the fibers of the inner and the tunnel spiral bundles. The nerve fibers forming this type of ending are small in diameter, usually about 0.2 μm. They form numerous synapses on afferent dendrites innervating the inner hair cells (Smith 1961; Smith and Rasmussen 1963; Iurato 1974). Dircet synaptic contacts between efferent elements and the inner hair cells are extremely rarely observed, therefore the axo-dendrite type of nerve ending (with the afferent fibers) is the only mechanism of efferent regulation in the inner hair receptors. This system is highly developed. For example, each afferent dendrite of the inner hair cell is known to possess up to 20 synaptic contacts with the efferent fibers of the inner spiral bundle (Liberman 1980). Axo-dendrite synapses have also been described in the fibers of the spiral bundle below the outer hair cells (Smith and Rasmussen 1963).

The hair cells in the organ of Corti are supplied by a branch of the auditory (VIII) nerve. In man and monkey, there are about 35 000 afferent fibers in the auditory nerve, in the guinea pig about 24 000, and about 50 000 in the cat. The nerve fibers are tonotopically organized in the auditory nerve according to the area of their innervation (Sando 1965). The auditory nerve contains afferent and efferent fibers. The number of efferent fibers does not exceed 5% of the total number of nerve fibers. It is interesting that 95% of the afferent fibers innervate inner hair cells and only 5% of the spiral ganglion fibers form synaptic connections with outer hair cells (Spoendlin 1972, 1985). It should be noted that the number of outer hair cells constitutes more than three-quarters of the receptor cells in the organ of Corti. This unusual pattern of innervation seems to be common for all mammals, including man. It seems important that the inner and the outer hair cells are innervated by both afferent and efferent fibers independently of each other.

1.2 Semicircular Canals

Semicircular canals in vertebrates are represented by three membrane structures forming membraneous rings which open at both ends into the utriculus. Their receptor apparatus is located at the ends of semicircular canals in special ampullae on whose cristae lies the sensory epithelium (Fig. 3). It consists of receptor and supporting cells as well as specialized cells. The latter are presumably involved in the production and secretion of endolymph. Receptor cells lie upon supporting ones. In cyclostomata, fish, and amphibia the receptor cells of the cristae ampullaris belong to the cells of type II; in reptiles, birds, and mammals there are two types of cells (Wersäll et al. 1965). Hair cells of type I are pear-shaped; their body, except for the apical membrane surface, is surrounded by a nerve chalice. The type II sensory cells are cylindrical and innervated by nerve terminals from small fibers, with several afferent and efferent nerve terminals. In mammalian cristae two thirds of the sensory cells belong to type I, the remainder to type II (Lindeman 1969). However, it has been recently distinguished four classes of isolated vestibular have cells from guinea pig (Scarfone et al. 1991).

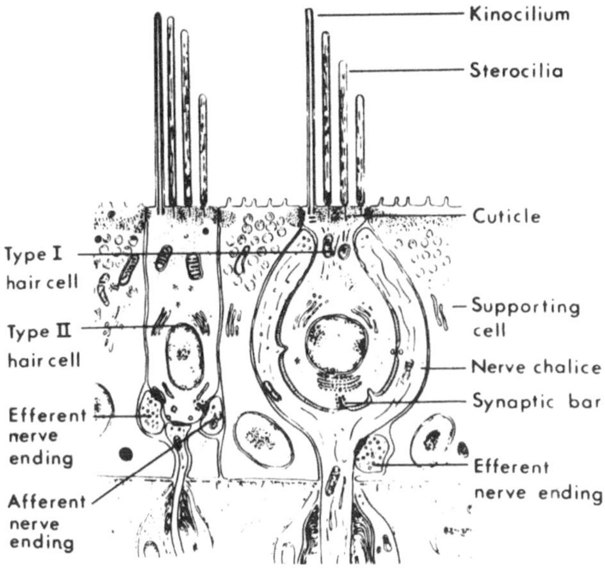

Fig. 3. Schematic drawing of the two types of sensory cells in the mammalian labyrinth showing the fine structure organization of the type I and the type II sensory cells and their innervation. (Wersäll and Bagger-Sjöbäck 1974)

The receptor cells carry on their apical surface 50–100 stereocilia and a single kinocilium located assymetrically to the bundle of stereocilia. The diameter of the kinocilium varies from 0.2 to 0.3 μm, their height amounts to 30–40 μm in the frog's ampullary cristae. Each kinocilium consists of nine peripheral and two central fibrils. In its basal part the kinocilium forms a small thickening – a neck, in which two central filaments end. At the same time, the peripheral pairs of fibrils penetrate into the cell, where the third, short filament joins them, forming a triple-tubuled cylinder. The peripheral fibrils end in the basal body of the cuticle.

The inner structure of the stereocilia differs from that of kinocilia. Their core consists of thin fibrils 40–60 Å in diameter, surrounded by a triple-layered plasma membrane. Stereocilia fuse towards their basal part; they penetrate into the cuticle and sometimes pass through the cuticle into the cytoplasm of the receptor cell. In most species stereocilia are 0.2 μm in diameter. Stereocilia are organized into hexagonal structures. The length of stereocilia gradually decreases as the distance from the kinocilium increases, so that the highest stereocilia are close to the kinocilium (Engström 1958). The excentric location of the kinocilium with respect to the stereocilia, and the latter's increase in length as they approach the kinocilium define a morphological polarization of the hair cell and are characteristic of receptor cells. Nevertheless, there may also be differences in the relative location of hairs and their length in different receptor areas (Lowenstein et al. 1968;

Wersäll and Bagger-Sjöbäck 1974). For example, in the cristae of the horizontal canal in each receptor cell the kinocilium is always oriented in the direction of the utriculus. In the receptor cells of the anterior and posterior vertical canals the opposite arrangement is observed.

An important role in the function of semicircular canals as the organ transducing angular acceleration belongs to the cupula. This gelatinous structure resting on top of the sensory epithelium is protruded into channels by the sensory hair bundles. During movement of the head or under the action of angular accelerations, as the result of the inertia of the endolymph, a displacement of the cupula and a shift in the endolymph relative to the walls of the membraneous canal can be observed. The movements of the cupula are transduced by the hair cells into excitation or inhibition of spike activity in afferent nerve fibers, depending on the direction of stimulation.

The vestibular sensory epithelium is supplied by the afferent nerve fibers of the vestibular ganglion cells. Efferent fibers run from nerve centers in the medula. The vestibular nerve is divided into an anterior and a posterior branch. The anterior branch innervates the anterior and horizontal canals, and also the macula utriculi, while the posterior branch of the vestibular nerve does so for the posterior canal and the macula sacculi.

The morphologic structure of synaptic contacts is typical of synapses with a chemical mechanism of transmitter release (Wersäll 1956; Wersäll et al. 1965; Bagger-Sjöbäck 1974). The synaptic structure in cyclostomes, fish, and amphibians appears fairly uniform. The cylindric receptor cells (of type II) and divided from the nerve endings by a synaptic cleft about 250 Å in diameter. In the zone adjacent to the presynaptic membrane, spherical or pear-shaped synaptic bodies can be observed. Each synaptic body is surrounded by numerous vesicles. In efferent synapses there are additional membranes creating a sort of cell cistern, while synaptic vesicles fill up the space of nerve terminals.

As pointed out above, in reptiles, birds, and mammals, besides cylindric hair cells there are cells of type I characterized by chalice synapses. The nerve endings of such synapses are filled with fine neurofilaments, mitochondria, and sparse vesicles whose number increases towards the apical end of the terminal. The chalice synapse contains several synaptic zones. This results in the fact that the synaptic space between the nerve chalice and the sensory cell of type I varies in width and structure. In the synaptic zones, the width of the synaptic cleft reaches 250 Å. In the rest of the contacts, the invagination of the terminal into the membrane of the hair cell is observed, synaptic space is greatly reduced, and in some contact zones a fusion of the plasma membrane of the sensory cells and that of the nerve chalice is observed (gap junction) (Spoendlin 1964; Hamilton 1968). In this connection it was suggested that the zones of close contact may be the sites of electric transmission, whereas the chalice synapse itself may be considered as a mixed synapse. However, there are no physiological data to support this hypothesis. Moreover, no gap junctions between hair receptor cells and afferent terminals have been described definitively in the organ of Corti and the semicircular canals. Earlier findings in this respect have not been confirmed by the freeze fracture technique (Gulley and Reese 1977; Luciano et al. 1977; Schessel et al. 1982).

1.3 The Otolith Organs

The receptor epithelium of otolith organs is represented by receptor and supporting cells (Fig. 4). In cyclostomes, fish, and amphibians, there are only cells of type II in receptor epithelium. In reptiles, birds, and mammals there are hair cells of type I as well. Cells of type I seem to be most developed in birds. Here, the number of these cells located in chalice endings may reach five to seven, whereas in mammals their number does not exceed two. Each cell in such a chalice has its own inclination with respect to other hair cells. This increase in the number of receptor cells in one ending, which renders the structure of their synaptic connections so complicated, may be conditioned by the necessity of precise movement coordination during the flight of birds (Vinnikov 1979).

The hair cells of the otolith organs, together with the afferent and efferent synapses, have general structures that are similar to those found in the cristae ampullaris (Smith 1956, 1967; Flock 1964; Vinnikov et al. 1965; Vinnikov 1979). Here the sensory cells are provided with stereocilia and a polarly located kinocilium (Fig. 5A,B). The latter is from 5–10 to 30–40 μm in diameter. It contains nine pairs of peripheral and two pairs of central

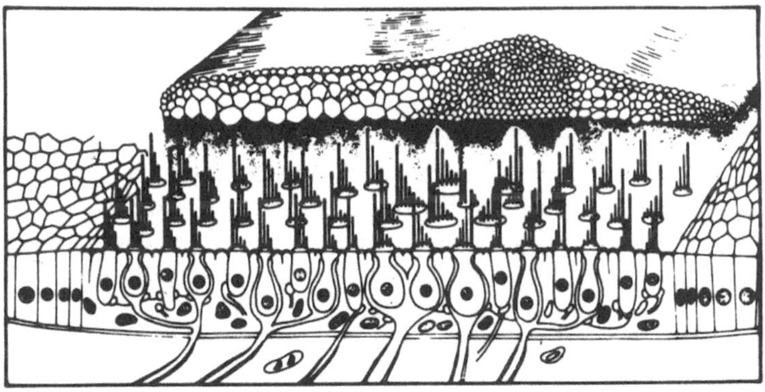

Fig. 4. Structural organization of receptor maculae sacculus of mammals. (Lindeman 1969)

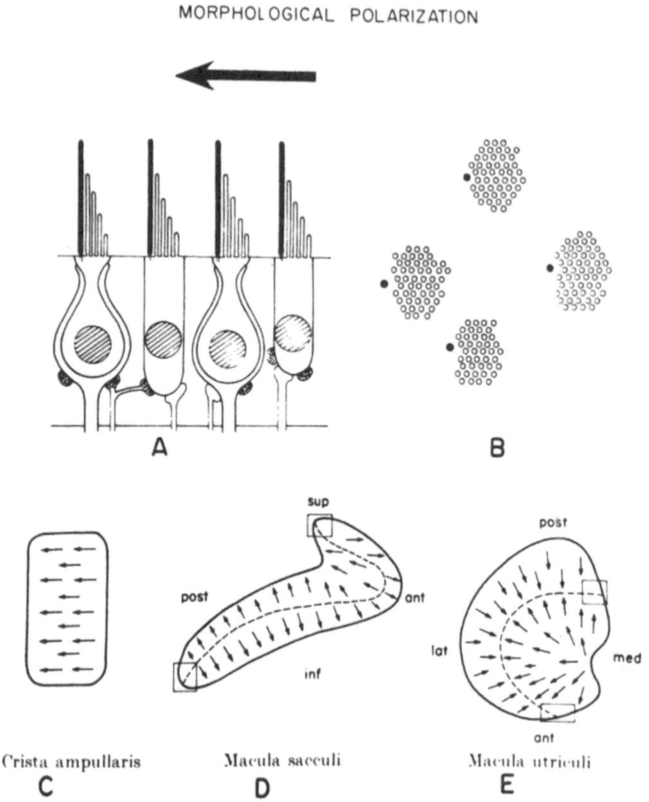

Fig. 5A–E. Morphological polarization of hair cells and directions of polarization of hair cells in otholitic organs of mammals (Lindeman 1969). **A** Schematical representation of the transverse section through hair cells. **B** Cross-section from hair bundles illustrating the regularity in the arrangement of the stereocilia and the kinocilia (black circles) located assymetrically to the bundle of stereocilia. **C–E** Schematical representation of receptor cells orientation in crista ampullaris, macula sacculi and macula utriculi

fibrils. On the outside, the kinocilium is covered with an extension of the receptor cell membrane. It should be noted, however, that under the scanning microscope six types of kinocilia were found in frogs, varying in length and in the presence of a thickening at the distal end (Lewis and Li 1975).

Stereocilia are somewhat thinner than kinocilia, 0.1–0.2 μm, and consist of a bundle of fibrils 20–40 Å in diameter. Stereocilia are usually arranged in regular rows. Each kinocilium is opposed by seven to nine rows of stereocilia. The length of stereocilia increases from 0.5 to 5 μm towards the kinocilium. In some cases one or two stereocilia are found under one plasma envelope (Vinnikov et al. 1971).

The area under the apical membrane of the receptor cell contains a granular cuticular plate. Only a small site at the apex of the cell is free;

here, the plasma membrane has direct contact with the cytoplasm. It is this cuticle-free site that contains the basal body of the kinocilium.

Each macula is covered by a thin extracellular fibrous structure, the otolith membrane. Layers of calcium carbonate crystals, the otoconia, are embedded in the upper surface of the membrane. Between the otolith membrane and the apical membrane of the hair cell there is a space in which the ciliary apparatus is located (Igarashi and Alford 1969; Dohlman 1971). The otolith membrane is permeated with channels into which the hair bundles enter.

In the ultrathin sections and under the scanning microscope it can be seen that, unlike stereocilia, which usually appear to be straightened, the kinocilium may occupy different positions (Vinnikov et al. 1971; Hillman 1972; Bagger-Sjöbäck 1974; Flock et al. 1977). This testifies to the ability of the kinocilium to move. It is suggested that because of its mobility the kinocilium returns the otolith membrane and the otoconia to their initial position (Vinnikov et al. 1971). In different species otoliths and otoconia differ in shape and size. For example, in fish, the otolith membrane of the utriculus, sacculus, and lagena is covered with a giant otolith, whereas in birds, on the outer surface of the otolith membrane there are many small otoconia. Owing to the inclusions of otoconia into the otolith membrane, the latter has a higher specific gravity than the surrounding endolymph. When the animal changes the position of its head in the gravitation field, and also during linear and radial accelerations, the gliding movements of otoliths occur and the "shearing" forces of the otolith membrane with respect to sensory epithelium are developed (Fig. 6). These forces cause the development of excitation and inhibition processes in different parts of the macula, in accordance with the complex pattern of the location and orientation of hair cells in the maculae.

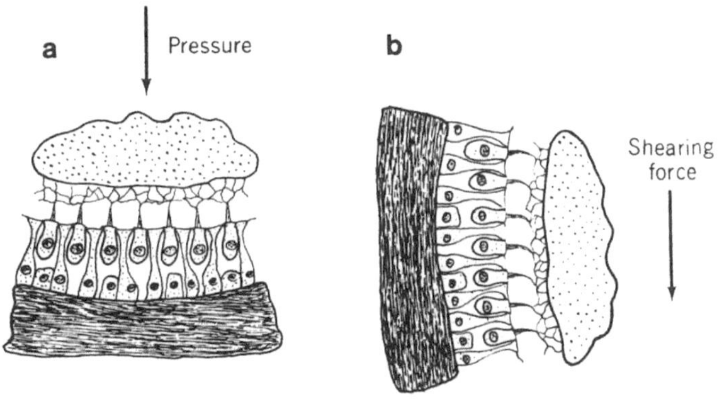

Fig. 6. Action of gravitational force on a macular ending, in two orientations (Wever 1974)

Receptor cells in the maculae of otolith organs are specifically oriented (Fig. 5C–E). For example, in teleosts and elasmobranchs (Flock 1964; Lowenstein et al. 1964) in the utriculus they are directed mainly in the shape of semicircles towards the anterior and the posterior edges of the macula, whereas in the boundary zone of the macula the hair cells are oriented in the opposite direction – their kinocilia are directed towards the kinocilia of the medial part. In the sacculus, the kinocilia of the hair cells are at right angles to the longitudinal axis of the macula. The kinocilia in the dorsal part of the sacculus are directed dorsally, those in the ventral part vertically, which results in the creation of two directions of receptor cell polarity in the sacculus. Data concerning cell orientation in the macula lagena in fish are lacking. In other species a still greater complication of hair cell polarization in the receptor fields of the utriculus, sacculus, and lagena can be observed in their evolutionary line (Vinnikov et al. 1971). There is reason to suppose that the complexity of reception field organization in otolith organs is directly connected with the development of the animals' skeletal musculature and the peculiarities of their movements in the Earth's gravitational field. In this respect, the extraordinary complication of hair cell polarization in the utriculus and sacculus of birds seems to reflect the adaptation of birds to flight.

The receptor cells of the otolith organs in their basal part possess structures typical of chemical transmitting synapses. Pear-shaped receptor cells (type I) synaptically come into contact with huge nerve endings which surround one or even several receptor cells on all sides. Only the apical membrane with the hair apparatus protrudes from the nerve chalice. In this case, practically the whole plasma membrane of the receptor cell may possibly be considered as a presynaptic one. However, the synaptic function is performed not by the whole cell membrane but only by its sites, where the distance between the pre- and postsynaptic membrane is shorter and additional synaptic structures in the shape of rod-like synaptic bodies appear (Wersäll and Bagger-Sjöbäck 1974).

Cylindric receptor cells (type II) form synaptic contacts on their basal surface with afferent and efferent nerve endings (Wersäll et al. 1965). Most such endings contain a great number of small mitochondria and sparse synaptic vesicles. The relations between nerve chalice and nerve endings of type II sensory cells and receptor cells may also be more complicated. For example, the nerve chalice may come in contact not only with cells of type I, but with the other side it may form typical synaptic contacts with cells of type II. Sometimes it possesses branches innervating the basal surfaces of cylindric cells and forming bouton-like nerve endings.

Besides afferent innervation, many investigators describe dark nerve terminals in otolith organs which approach receptor cells. Such nerve endings, containing a great number of synaptic vesicles and a few mitochondria, seem to be efferent. They approach both the basal surface of the type II sensory cells and form axo-axonal synapses with nerve endings surrounding receptor cells of type I.

1.4 Lateral-Line Organs

The lateral-line organs are epidermal sense organs found in cyclostomes, fish, larval amphibians, aquatic adult urodeles, and some adult aquatic anurans. The two main types of mechanosensitive lateral-line organs or neuromasts are the free-standing superficial organs and the canal organs (Fig. 7). In teleosts and amphibians, the neuromast consists of 20–60 receptor cells supplied with cilia, or hairs, the supporting cells separating and surrounding them, and a gelatinous cupula covering the hair cells from above. The accumulation of receptor and supporting cells forms the sensory epithelium. In each neuromast, the supporting cells are connected to the

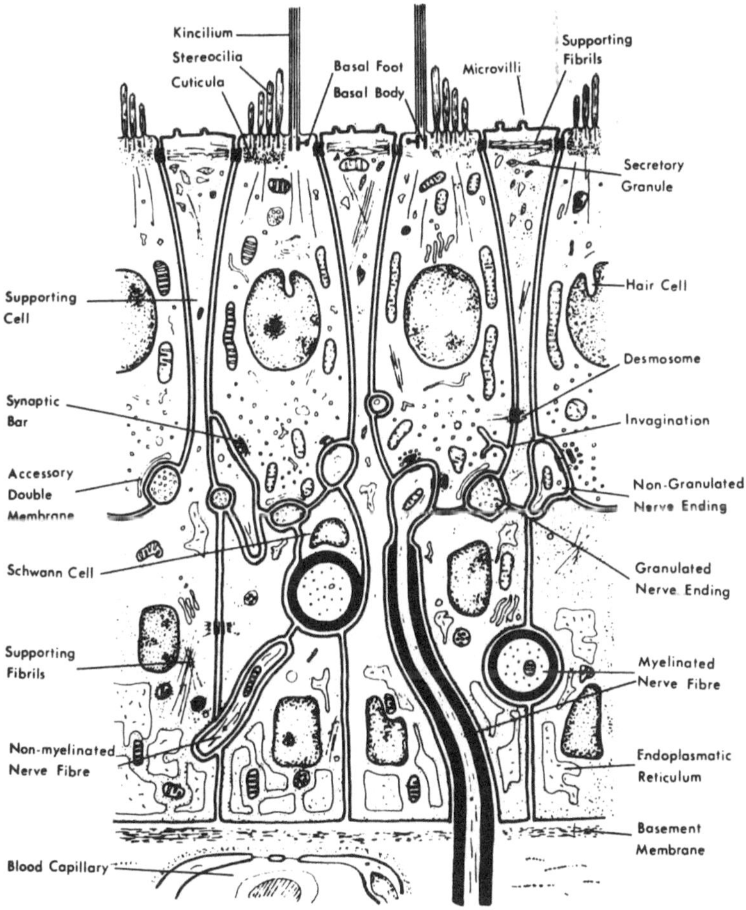

Fig. 7. Schematic drawing showing the ultrastructure of the sensory epithelium of the lateral line canal organs in *Lota vulgaris*. (Flock 1965)

hair cells through a system of desmosomes. Cylindrical hair cells are 30–40 μm in length and 7–10 μm in width. Their apical surface has a bundle of stereocilia and a single kinocilium about 40 μm in length (Flock 1965). With their distal ends, stereocilia and the kinocilium project into the cupula which may extend up to 100 μm into the surrounding water in free-standing neuromasts (Görner 1963). Since the kinocilium is always located asymmetrically with respect to the bundle of stereocilia, the receptor cell appears to be morphologically polarized.

Each stereocilium (0.2 μm in diameter) consists of a fibrillar core surrounded by a membrane, which turns into a plasma membrane at the base of the hair. In the center of the stereocilium there is a bundle of fibrils ending in roots in the granular cuticular plate of the cell. Kinocilia, unlike stereocilia, have nine pairs of fibrils in their core, encircling one central pair of fibrils. The central pair of fibrils appears to be oriented perpendicularly to the rows of stereocilia. Peripheral fibrils terminate on the basal body located at the base of the kinocilium. In elasmobranchs, electronmicroscopic investigation revealed two types of receptor cells (Roberts and Ryan 1971). Cells of type I are similar to the hair cells of teleosts and amphibians. Cells of type II are devoid of stereocilia but their apical surface carries a group of microvilli without inner fibrils. Whether morphological differences in hair cells of elasmobranchs are connected with some functional peculiarities is still obscure.

The receptor cells of the lateral-line organs, as well as almost all the hair cells of the acousticolateral system, are characterized by the presence of special crystalline inclusions in their cytoplasm and even in their nucleus (Mullinger 1964; Yamada and Hama 1972; Wickham 1973; Toyoshima and Shimamura 1982). They have no surrounding membrane and are represented by parallel electron-dense laminar structures about 25 μm in diameter. The functional role of the crystalline inclusions remains obscure. However, the fact that they are essential components of the cytoplasma of all mechanoreceptor elements, and that they are absent in the supporting cells and some other receptor structures which do not belong to the acousticolateral system, may indicate their involvement in the reception of mechanical stimuli by hair cells.

Relatively little is known about the structure and function of supporting cells. The supporting cells are usually longer than the receptor cells, and are also supplied with cilia on their apical membrane. The cell cytoplasm contains granules 0.1–0.3 μm in diameter, microfilaments, and also the usual set of cell organoids, especially in its apical part. There is strong evidence suggesting that in the lateral-line organs the supporting cells are involved in the secretory process of a gelatinous cupula covering the hair cells from above (Cornford and Barajas 1977).

A single neuromast is innervated by a thin bundle of fibers of the lateral-line nerve. In amphibians, it usually contains three or four myelinated nerve fibers which are accompanied by a bundle of nonmyelinated fibers

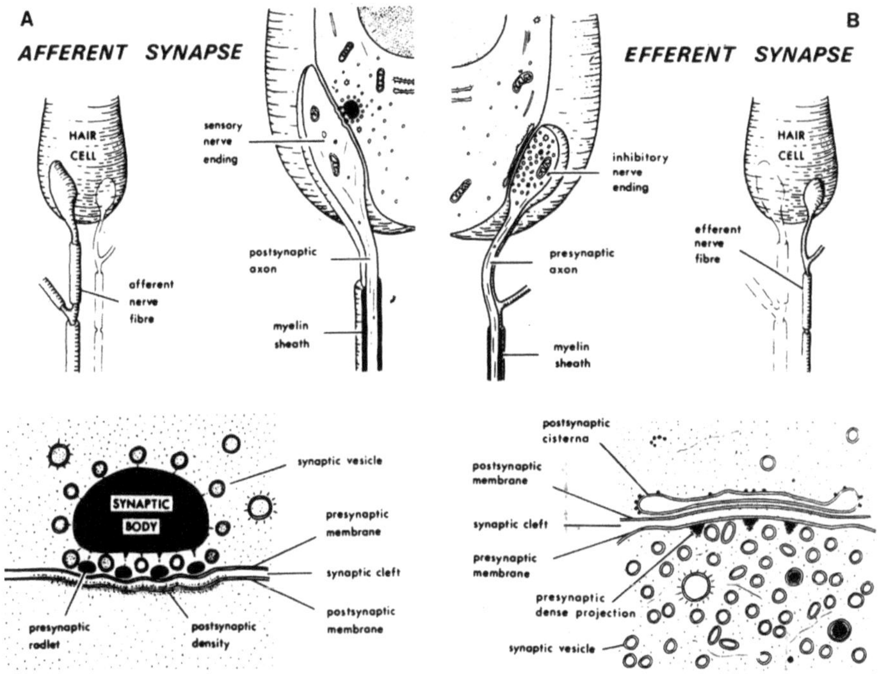

Fig. 8A,B. Anatomy of afferent (**A**) and efferent (**B**) synapses in hair cells of lateral-line organs. (Benshalom and Flock 1980)

(Jørgensen and Flock 1973). When approaching the neuromast, which in amphibians consists of approximately eight cells, the fibers lose their myeline sheath, divide into a number of terminals, and supply the sensory cells. In fish, a single neuromast containing 10–50 hair cells is usually innervated by several nerve fibers. At the basal part of the hair cells granular and non-granular endings are seen (Flock 1965; Hama 1965) (Fig. 8).

Endings of type I, efferent, are characterized by a great number of vesicles (0.03–0.04 μm in diameter) in the synaptic area of the cell and nerve membranes. The efferent function can be performed both by fine-myelinated (Görner 1963; Russell 1968) and nonmyelinated (Harris and Flock 1967) fibers. In separate receptor cells of the neuromasts efferent innervation may be absent (Hama 1965).

Endings of type II are afferent, their nerve membrane covers a considerable surface of the basal part of the cell, forming a synaptic cleft 0.015–0.02 μm in width. In the afferent synapse, pre- and postsynaptic membranes are separated by a synaptic cleft about 150 Å in diameter which looks more electron-dense than the rest of the extracellular space between the hair cell and the nerve ending. In the synaptic cleft dark filaments can be seen connecting pre- and postsynaptic membranes and are probably

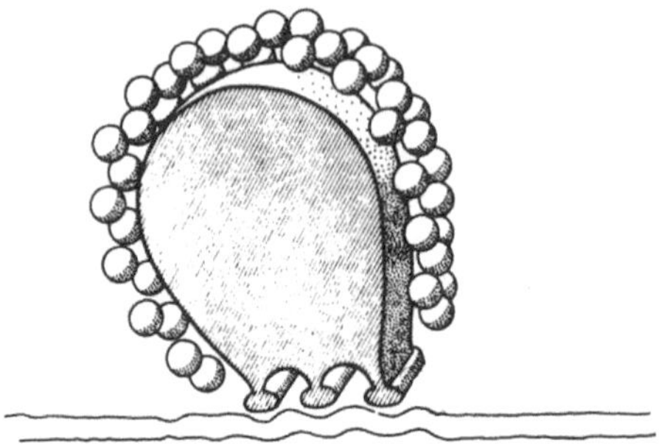

Fig. 9. Schematic drawing of the synaptic body in frog. (Gleisner et al. 1973)

responsible for the maintenance of the constant distance between the membranes in the area of synaptic contacts (Benshalom and Flock 1980). The pre- and postsynaptic membranes are parallel to each other although they look somewhat "wavy". The postsynaptic membrane is somewhat thicker than the presynaptic one. The thickness of the presynaptic membrane does not differ from that of the rest of the hair cell membrane.

The hair cells of the acousticolateral system are distinguished by the presence of the so-called synaptic bodies in their cytoplasm (Figs. 8A, 9). They are described for the lateral-line organs in fish (Flock 1965; Hama 1965; Benshalom and Flock 1980) and amphibians (Shelton 1971; Jørgensen and Flock 1973), in the electroreceptors of nonelectric and weakly electric fish (Mullinger 1964, 1969; Szabo and Wersäll 1970; Szamier and Wachtel 1970), in the ampullae of Lorenzini (Barets and Szabo 1962; Waltman 1966; Derbin 1970; Jørgensen et al. 1972; Ilyinsky et al. 1974), in the organ of Corti (Smith and Sjöstrand 1961; Gleisner et al. 1973; Saito 1980) (Fig. 8), and in the vestibular apparatus (Hama and Saito 1977; Favre and Sans 1979). Synaptic bodies are electron-dense formations of variable sizes and shapes (ribbons, bars, sheets, etc.). They are found in the regions of hair cells directly adjacent to the presynaptic membrane. Detailed studies of the structure of hair cells in the canal organs of the fish have shown that the synaptic body consists of two subunits: the main body mass and its feet (Benshalom and Flock 1980). The feet of the synaptic body are elongated formations fitting into the grooves of the presynaptic membrane. One synaptic body is connected to about 14 feet. In cross-section they appear as round bodies, whereas in longitudinal sections one can see that they consist of microtubules. The main body mass is connected to the feet by thin filaments. Around synaptic bodies there is a considerable concentration of synaptic vesicles 30 nm in diameter without an electron-dense core. Vesicles with a dense core and coated vesicles can be observed in other parts of the cytoplasm.

The functional role of synaptic bodies is still obscure. But their presence at afferent synapses in sense organs makes it possible to suppose that in all cases they perform a similar function. It is suggested that synaptic bodies may be a site for the storage of neurotransmitters or their precursers in the cell (Osborne and Thornhill 1972; Wagner 1973; Monaghan 1975; Osborne 1977). It should be noted that acoustic stimulation of the frog's basilar papilla causes an increase in the size of the synaptic bodies. In lateral-line organs of salamander tadpoles, where synaptic bodies can be seen in vivo in the transparent tail, they are seen to move to the cell base during receptor stimulation (Flock et al. 1973a; Flock 1974). These data unambiguously point to the role of synaptic bodies in the plasticity function of sensory synapses. Synaptic bodies are also supposed to control the guidance and access of synaptic vesicles to the presynaptic membrane (Benshalom and Flock 1980); for example, when continuously stimulated, synaptic bodies can enhance the transmitting ability of the afferent synapse. At the same time, their size and number increase. Under conditions of relative rest, a reverse process occurs. The presumed role of synaptic bodies as sites for calcium binding was rejected as a result of investigation (Benshalom and Flock 1977).

In the efferent synapse, the vesicles are concentrated in the nerve ending. The presynaptic membrane of the nerve fiber does not differ in thickness from the rest of the membrane but contains characteristic electron-dense projections 30 nm wide and 20–30 nm high. The postsynaptic membrane is not more electron-dense than the rest of the hair cell membrane. Pre- and postsynaptic membranes run parallel to each other, apparently due to intrasynaptic filaments. The efferent synapses possess special membrane structures, referred to as postsynaptic cisterns, in the area adjacent to the inner side of the postsynaptic membrane (Benshalom and Flock 1980). These membrane formations may be elongated and are adjacent to the postsynaptic membrane at a distance considerably smaller than the width of the synaptic cleft. The stability of the distance between the postsynaptic membrane and the cistern membrane is maintained by electron-dense material. The cistern membrane facing the inner part of the hair cell usually carries ribosome-like granules on its surface, usually grouped in threes.

The lateral-line organs are inervated by the anterior (n.lat.ant.) and posterior (n.lat.post.) lateral-line nerves. The anterior lateral-line nerve supplies receptor structures on the head, whereas the posterior lateral-line nerve does so for the body of the animal. Lateral-line nerves are peripheral projections of ganglion cells located intercranially. The anterior lateral-line nerve is formed by a branch of the facial (VII) or trigeminal (V) nerve, whereas the posterior one is formed by the branches of the vagus (X) and the glossopharyngeal (IX) nerves. Therefore, the anterior nerve is sometimes called n.lat.facialis, and the posterior one n.lat.vagi (Ariëns Kappers et al. 1936).

As noted above, two main types of mechanoreceptor structure of the lateral line are known: the free-standing neuromasts and the canal organs. The main differences between them are the peculiarities of their accessory apparatus structure, which allows to them to come into direct contact with the environment. The canal organs have no such contact, and the external stimulus primarily causes the displacement of canal fluid which activates receptors. In teleosts, the sensory canals consist of five main branches, the main lateral canal, the supratemporal, temporal, and supra- and infraorbital canals.

Considerably different is the system of canal organs in bottom-dwelling fish with their flattened bodies. In skates, for example, in the process of evolution thorax fins grew into the head and widened, which was accompanied by deep changes in structure of the lateral-line system, primarily concerning the position of the canal organs of the lateral and lower parts of the head. In teleosts or sharks, these are located in the lateral part of the head, whereas in skates they are very elongated and located mainly on the upper and lower parts of the fins (Disler 1960). The peculiarities of canal topography in the body of the fish correlate with the direction of the water flow during movement (Tester and Kendall 1969).

The sensory canals of the lateral line are filled with a viscous fluid whose ion composition differs from that of the cerebrospinal fluid and blood plasma (Ilyinsky and Krasnikova 1971; Ilyinsky 1975). Available data suggest an elevated level of potassium content in the canals and a content of sodium, calcium, magnesium, and chloride ions close to that of the cerebrospinal fluid and blood plasma. An elevated level of potassium ion content is also characteristic of other highly specialized mechanoreceptors and primarily of cochlear and vestibular receptors. This fact seems to be another proof of the existence of a phylogenetic connection between the structures of the lateral-line organs and the inner ear.

Little evidence exists concerning the content of organic substances in the canal fluid. Thus, it is known that there is a considerably smaller amount of protein in the fluid than in the blood plasma or the cerebrospinal fluid. At the same time, the content of mucopolysaccharides may attain high levels as, for example, observed in the skate canal fluid (about 21%) (Ilyinsky 1975). Owing to the ability of acid mucopolysaccharides to bind different cations, a considerable part of the potassium ions (21.9% in the skate canal fluid) seems to be bound in the canal fluid.

1.5 Electroreceptors

As noted above, electroreceptors are specialized receptors of the lateral-line organs. The problem of electroreceptor classification is not yet solved, but Szabo's classification (1965, 1974) may be considered the best. According to this, all electroreceptors are divided into ampullary and tuberous ones. The main morphological difference between them is that in the ampullary organs the apical sensory-cell membrane is connected by a jellyfield canal to the epidermal surface, whereas tuberous receptors do not possess such a canal and the sensory cells are connected to the surface by way of epidermal cells covering the sense organ. Ampullary receptors occur in most nonteleosts, in Siluriformes, Mormyriformes, Gymnotiformes, in a subfamily of Osteoglossomorpha, in Xenomystinae, and in some amphibians. Tuberous receptors are observed only in weakly electric fish. Recently, a type of tuberous receptor was found in one species of Siluriformes, *Pseudocetopsis* sp. (Andres et al. 1988).

1.5.1 Ampullary Receptors

The ampullary receptors include the ampullary organs of weakly electric fish and of some amphibians, the ampullae of Lorenzini of nonteleost fishes and of the only marine teleost, *Plotosus*, and the small pit organs of the catfish. In principle, all the ampullary organs of the catfish and of weakly electric fish are similar in structure. An isolated organ consists of a canal connecting the sensory cells to the skin surface and ending in a blind cavity – the ampulla, in the base of which lies the sensory epithelium. The ampullary organs are distributed over the whole dermis in teleosts, but in the head their concentration is especially high (Peters et al. 1974; Jakubovski 1987). The length of the canal varies in different species. For example, in weakly electric fish it is 100–200 µm, in catfish the canal is considerably shorter, and in the transparent catfish *Kryptopterus* it is practically absent, so that the ampullae open directly on the skin surface (Mullinger 1964; Szamier and Wachtel 1969; Wachtel and Szamier 1969). The canal wall consists of three to six cell layers linked together by desmosomes in the outer layers or by tight junctions in the inner layers (Lissmann and Mullinger 1968; Szamier and Wachtel 1969, 1970; Wachtel and Szamier 1969).

The sensory epithelium is composed of two types of cells, receptor and supporting cells. Receptor cells, 10–15 µm in diameter, are separated from each other by the supporting cells which surround them on all sides, leaving free only a small (about 15%) part of the apical cell membrane which opens into the canal lumen. From the apical membrane, numerous microvilli, 0.8 µm in length and 0.1 µm in diameter, project into the lumen but no ciliary apparatus characteristic of other receptor cells in the acousticolateral system is found here (Roth and Tscharntke 1976; Jørgensen 1980, 1982; Sejnowski and Yodlowski 1982; Jakubowski 1987). The function of microvilli is still obscure. Besides ampullary organs, they were also found in a great number of different epithelial cells. It cannot be ruled out that their role is confined to increasing the apical membrane surface area.

The receptor cells, 10–15 µm in diameter, are oval and possess a relatively large round nucleus about 5 µm in diameter. In the region of the presynaptic membrane there is a synaptic body, 1–2 µm in diameter, whose shape varies from species to species. The synaptic cleft is about 20 nm wide and rather long. Single fibers about 5 μm in diameter end in large terminal boutons on each receptor cell, which forms about 20–30 synapses with each bouton (Szamier and Wachtel 1969, 1970). This considerably increases the surface area of synaptic contacts.

The number of receptor cells in each ampullary organ is very small compared to the ampullae of Lorenzini. Thus, in gymnarchids there is only one receptor cell per microampulla, in mormyrids and gymnotids there are four to eight cells, in the transparent catfish their number varies from eight to 20, and in the European wels each ampulla contains about eight receptor cells (Lissman and Mullinger 1968; Szamier and Wachtel 1969, 1970; Szabo 1974; Jakubowski 1987). In most teleosts each ampulla is innervated by a single nerve fiber which comes in contact with each receptor cell in the ampullary organ.

The ampullae of Lorenzini are organized somewhat differently. The length of the canal in these organs, correlating with marine versus freshwater habitat, varies from 500–800 μm in the freshwater skate (Szabo et al. 1972; Szamier and Bennett 1980) to several dozen cm in large marine skates (Fig. 10). The canals are located under the skin and are of different length

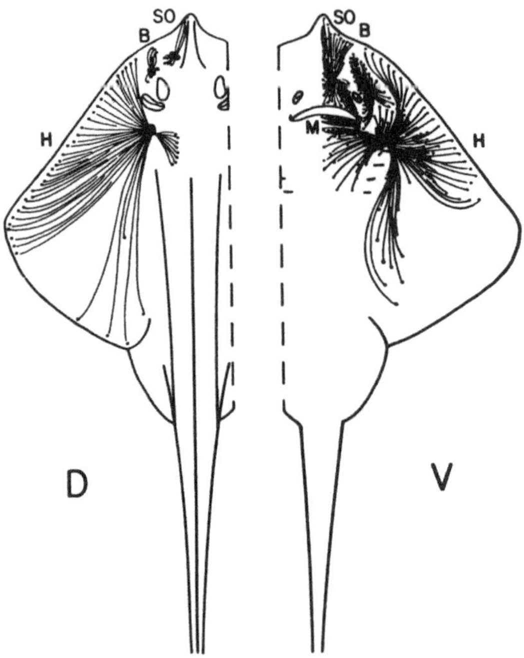

Fig. 10. Distribution of the electroreceptor organs over the dorsal (*left*) and ventral (*right*) surface of the skate *Raja clavata*. The ampullae of Lorenzini occur in four groups. The ampullae proper of each group are contained in a common capsule and each of the capsules is innervated by a separate ramus of anterior lateral line nerve: superficial opthalmic (*SO*), buccal (*B*), hyoid (*H*), and mandibular (*M*). (Murray 1965)

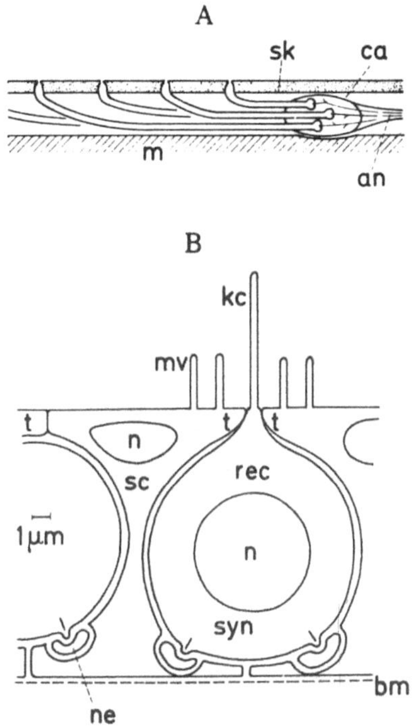

Fig. 11A,B. Diagrams illustrating the anatomy of the ampullae of Lorenzini. **A** Section through skin showing canals ending in ampullae within a capsule. *sk* skin (epidermal + dermis); *ca* capsule; *an* ampullary nerve; *m* body muscles. (Murray 1974). **B** Section through the sensory epithelium showing the receptor and supporting cells. *bm* basement membrane; *kc* kinocilium; *mv* microvilli; *n* nucleus; *ne* nerve ending; *rec* receptor cell; *sc* supporting cell; *t* tight junction; *syn* synapse. (Waltman 1966)

since they originate at different points of the body and converge in capsules where they terminate in blind ends or ampullae. On the bottom of them are receptor and supporting cells. The canal wall is composed of a double layer of flattened cells, with tight junctions between the cells of the inner layer and desmosomes between the cells of the outer layer (Waltman 1966).

The ampullae of Lorenzini of different elasmobranchs and *Plotosus* are fairly similar to each other in structure and differ only in some details of the organization of the ampulla itself (Waltman 1966; Jørgensen et al. 1972; Szabo 1972; Ilyinsky et al. 1974; Zakon 1986; Raschi 1986; Raschi and Mackanos 1987) (Fig. 11). In marine species each ampulla contains several thousands of receptor cells; in the ampullae of freshwater skates there are about 50 cells. Receptor cells are separated from each other and surrounded by the supporting cells so that only about 1% of the total membrane faces into the lumen in marine elasmobranchs. The apical surface of each receptor cell is ringed with tight junctions to the surrounding supporting cells (Waltman 1966; Sejnowski and Yodlowski 1982). This continuous network of tight junctions on the surface of the sensory epithelium effectively attenuates any current leakage through intercellular clefts and the canal walls and creates the basis for electroreceptor function. Most of the cell membrane underlying the tight junction may be considered the basal membrane. Thus, the existence of tight junctions results in the apical and basal membranes being electrically isolated from each other and facing fluids that differ in ionic composition. A similar situation seems to be observed in other hair cells of the acousticolateral system (Flock 1967; Hudspeth 1983; Zakon 1986).

The receptor cells in the ampullae of Lorenzini are pear-shaped, 10–15 μm in diameter. The apical surface of receptor cells in elasmobranchs and chondreosteans bears a single kinocilium protruding into the ampullary lume. In *Plotosus* the sensory cells have no kinocilium but bear long microvilli, 200–300 in number, but well arranged like stereocilia, in several rows. In polypteriform fish, the lungfish (Roth and Tscharntke 1976; Zakon 1986), and apodan amphibians (Fritzsch et al. 1985; Fritzsch and Munz 1986), there are both microvilli and a kinocilium, whereas in salamander larvae there are only microvilli (Fritzsch and Wahnschaffe 1983). The kinocilium is about 0.3 μm in diameter. In its inner core it contains 8 + 1 pairs of fibrils, which reverts to 9 + 0 pairs at the base of the kinocilium. As with other kinocilia, it terminates on the basal body. It remains unclear whether the kinocilium has any functional role in electroreceptors or its existence proves that electroreceptors may have developed from ordinary lateral-line organs (free-standing neuromasts or canal organs) (Lissmann and Mullinger 1968; Bennett 1971; Jørgensen 1982).

Each ampulla of Lorenzini is innervated by a bundle of nerve fibers. Thus, every receptor cell whose number reaches several thousands is contacted by several nerve fibers, and one fiber innervates hundreds and sometimes thousands of receptor cells (Barets and Szabo 1962; Waltman 1966; Sejnowski and Yodlowski 1982; Fields and Ellisman 1985; Zakon 1986). Such convergence of receptor cells is very important for electroreceptor functioning. It increases the signal-to-noise ratio and thereby enhances the sensitivity of the electroreceptor organ to electric stimuli.

Pre- and postsynaptic membranes in the receptor elements of the ampullae of Lorenzini are separated by a synaptic cleft about 2 μm wide, which suggests the existence of a chemical transmission at this synapse. The presynaptic membrane forms a characteristic finger-like process and enters the invagination of nerve ending. Each receptor cell makes from 20 to 30 such synaptic contacts with afferent nerve terminals. In the cytoplasm of the receptor cell there is a synaptic body close to the inner surface of the presynaptic membrane surrounded with synaptic vesicles (Fig. 12). In the ampullae of Lorenzini, synaptic bodies have a characteristic ribbon structure, therefore the synapses are sometimes referred to as ribbon synapses. Recently, a correlation has been found between the morphology of the ribbon synapse and the sensitivity of the organ to electric stimulus (Fields

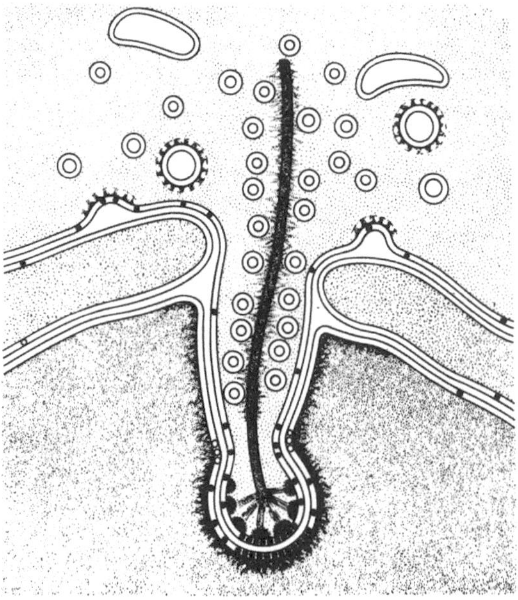

Fig. 12. Schematic drawing of ribbon synapse in the ampullae of Lorenzini. (Sejnowski and Yodlowski 1982)

and Ellisman 1985). The depth of projection of the presynaptic membrane into the nerve terminal appeared to increase receptor sensitivity significantly. These data suggest the existence of structural and functional plasticity in such and similar synapses, and the interaction of these synaptic properties seems to determine such processes as habituation, sensitization, or long-term memory. No efferent innervation has been found in the ampullae of Lorenzini or in other types of electroreceptors.

The ampullary canals are filled with a gelatinous substance mainly consisting of mucopolysaccharides (Ilyinsky 1975). The ion composition of the jelly has been studied in the ampullae of Lorenzini of skates (Murray and Potts 1961; Ilyinsky and Krasnikova 1971, 1974) and the marine catfish *Plotosus* (Okitsu et al. 1978). It varies, depending upon the ion composition of the seawater which the animal inhabits, and is characterized by an elevated level of potassium ion content as compared to seawater (1.3–3.4 times higher than in skates and six times higher than in the catfish) and to cerebrospinal fluid (two to five times higher in skates and approximately 15 times higher in the catfish). The concentration of potassium ions is higher near the ampullae than near the pore opening. These findings suggest a secretion of potassium ions in the vicinity of the receptor cells.

Thus, the ampullary receptors function within some specific medium distinguished for the elevated level of potassium ions as compared to the normal extracellular levels. A similar situation is observed in the endolymph of the inner ear. A general tendency for maintaining a medium of specific composition and properties around the hair cells serves as still another proof of the existence of a phylogenetic connection between the inner ear receptors and the lateral-line organs and enhances their sensitivity to mechanical and electrical stimuli (Ilyinsky 1975; Zakon 1986).

Of great importance for detecting weak electrical signals in the aquatic environment are the electrical properties of the ampullary jelly and canal walls. These properties have been studied for the elasmobranch ampullae of Lorenzini (Waltman 1966). The average mean values of electrical characteristics for the ampullary canal with a diameter of 0.116 cm are:

- mural resistance 8.4 Ω;
- mural resistance x unit area 6.0 Ω/cm;
- canal space constant 74 cm;
- capacity 0.31 μF;
- mural capacity per unit area 0.43 μF/cm;
- mural time costant 2.5 s;
- core resistivity 30.5 Ω/cm.

These peculiarities of the ampullary canal impart to it the properties of an ideal submarine cable. This results in a negligible decrease in voltage along the canal. At the same time, owing to a fairly high time constant, the canal possesses the property of a low-pass filter, which makes the ampullary receptors more suitable for the analysis of relatively low-frequency electrical signals.

1.5.2 Tuberous Receptors

1.5.2.1 Common Tuberous Receptors

Tuberous receptors are located in the invaginations of the epidermal basement membrane (Fig. 13). Some organs contain receptor cells located in the epidermal capsule without coming into direct contact with the outer medium, since the interior of the canal is filled with plug cells connecting the capsule cavity with the skin surface (Szabo 1965, 1974; Zakon 1986). Receptor cells are attached to a hillock of supporting cells, whereas up to 90% of the receptor cell surface is exposed freely to the sensory cavity. The orientation of the receptor cells within the capsule may be quite different in various species. For example, in some gymnotids with a great number of receptor cells they are oriented with their apexes towards the center of the organ. On the contrary, in other tuberous organs, the bases of the receptor cells are directed radially to the center of the supporting cell hillock and, consequently, the cell apexes are directed centrifugally. The tuberous organ is supplied by a single nerve fiber which penetrates the basement

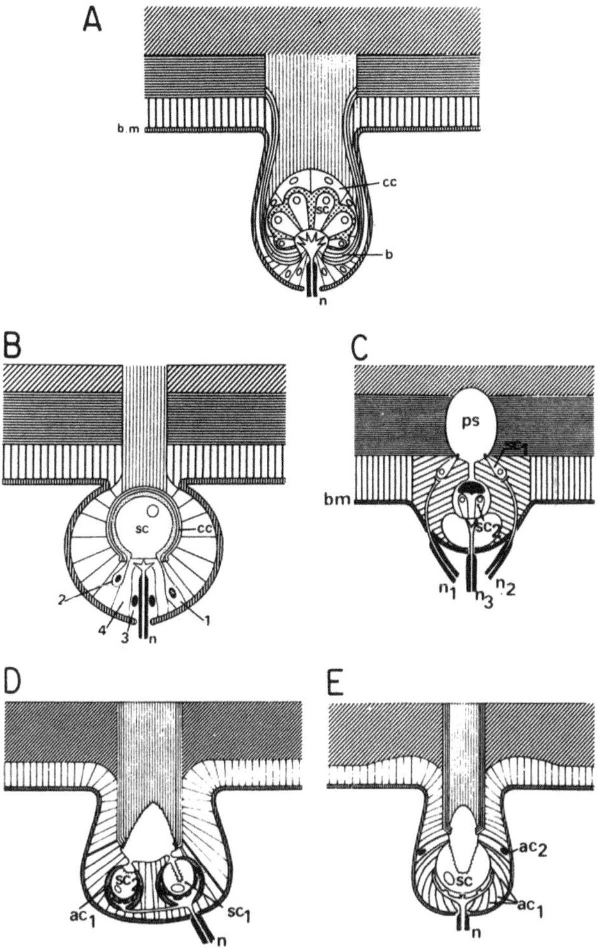

Fig. 13A–E. Schematic tuberous receptor organs. **A** Gymnotiform receptor organ. **B** Mormyrid knollenorgan. **C** Mormyrid mormyromast. **D** Gymnarchomast type I. **E** Gymnarchomast type II. *ac* accessory cells; *b* capsule wall; *bm* basement membrane; *cc* covering cells; *n* nerve terminals; *ps* perisensory space; *sc* sensory cells. *Numbers* refer to different types. (Szabo 1974)

membrane and, after branching between supporting cells, comes in contact with receptor cells.

The number of receptor cells in the capsule varies from 1 to 35 in mormyrids and from 10 to 100 in gymnotids (Szabo 1965, 1974). In the latter, the type I and II organs are distinguished according to the number of receptor cells. The type I organs usually contain about 20–30 receptor cells. The type II organs may have up to 100 cells. A typical characteristic of the tuberous organs in mormyrids is the location of each receptor cell in its own cavity, whereas in gymnotids, in similar organs all receptor cells occupy one common cavity. As pointed out above, the tuberous organs, as distinguished from ampullary ones, have no direct communication with the outer medium, since the space of the canals is filled by the so-called plug cells. They are cells of an epithelial nature coming in contact with each other with occasional filaments and desmosomes. The extracellular spaces between the plug cells seem to possess a low resistance pathway in the interior of the plug (Wachtel and Szamier 1966; Szabo 1974; Zakon 1986).

The receptor cells of tuberous receptors in gymnotids are elongated, 30 × 10 µm in diameter. In mormyrids, the cells are oval, 40 µm in diameter. In receptor cells, as in teleost ampullary receptor cells the ciliary apparatus is absent but on the apical surface there are a great number of microvilli 0.7–1 µm high and 0.15 µm in diameter extending radially into the perisensory space. Their density and their general number in tuberous organs are much higher than those on ampullary receptors, mainly due to the greater surface area of the tuberous receptor's apical membrane. The microvilli increase the total surface area of the apical membrane, which thereby increases its capacity and reduces its input resistance (Bennett 1967). In the hair cells of the acousticolateral system, stereocilia are known to play an important role in the processes of mechanoelectric transformation (Hudspeth 1983). However, it is difficult to conceive what functional role microvilli play in tuberous electroreceptors, and in other electroreceptors in general, even if only for the reason that in electroreceptors there are no such energy transductions because of their extremely low mechanical sensitivity.

With the aid of the freeze fracture technique it was possible to clarify the extracellular fluid compartments in the current pathways between the exterior and the sensory epithelium in the tuberous organs of gymnotids (Bennett et al. 1989). In these organs gap (communicating) and tight (occluding) junctions were identified that allowed the formation of an equivalent electric circuit of the receptor organ (Fig. 14).

The cells contain the usual set of cellular elements and are characterized by the presence in the basal part of the cytoplasm, and especially in the synaptic area, of a great number of large (0.05 µm) and small (0.03 µm) vesicles grouped around the synaptic body. In mormyrids, only a small part of the basal membrane are in synaptic contact with the ending of the nerve fiber, whereas the rest of the basal membrane contacts the supporting cells. In gymnotids a considerable part of the basal membrane surface comes in synaptic contact with a single bouton-like ending of the afferent nerve fiber. The ultrastructure of tuberous receptor synapses differs little from that of ampullary organs; only the terminal boutons are smaller. The only exception was found in a high-frequency gymnotid, *Sternarchus*, in which in the tuberous organs, alongside the structures typical of the chemical synapse, there are areas of close (less than 3 nm) oppositions of areas of the pre- and postsynaptic membranes: these data suggest the presence of electrical transmission in the synapse (Srivastava 1972). However, these findings have not been confirmed by the freeze fracture technique, which is more reliable for

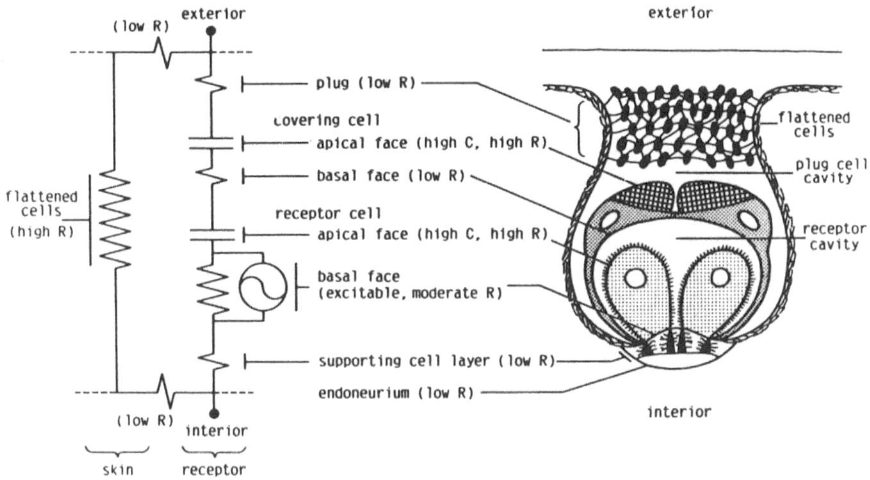

Fig. 14. Diagram of the receptor and its probable electrical equivalent circuit in high-frequency gymnotiform fish. (Bennett et al. 1989)

the identification of junctional relations (Bennett et al. 1989). Only in poorly fixed material could close appositions be seen in this region, but they were probably a fixation artifact. The negative results with this technique suggest that close appositions of pre- and postsynaptic membranes between receptor cells and afferent fibers in tuberous organs of gymnotids cannot be regarded as gap junctions and that the interpretation of Srivastava (1972) was in error. No efferent innervation in tuberous receptors has been found.

Innervation of tuberous organs in mormyrids is carried out by myelinated nerve fibers, whereas in gymnotids there are receptors with both myelinated and nonmyelinated innervation. It should be noted that although the innervation pattern in fact reflects an important functional aspect of sense organs, yet the study of this problem for tuberous organs in weakly electric fish has not been given proper attention so far.

1.5.2.2 Specific Tuberous Receptors

In addition to the common tuberous organs described above, a special group is constituted by specific tuberous organs, the so-called mormyromasts found only in mormyrids, and gymnarchomasts found in gymnarchids (Szabo 1965, 1974). In the mormyromast there are two types of sensory cells. Some of them, five to seven in number (type I) are located around the main intraepidermal cavity: others (type II) three to five in number, are located deeper in another intraepidermal cavity called the sensory chamber. Both cavities are connected by a canal.

Receptor cells of type I are surrounded by supporting cells, so that only a small part of the apical membrane is in contact with the intraepidermal cavity. Cells of type I do not possess any microvilli on the apical membrane. The basal cytoplasm forms synaptic contacts with numerous small nerve endings. In the area of the presynaptic membrane there is a synaptic ribbon-like body surrounded with synaptic vesicles. It is characteristic that in mormyromasts normal and degenerated sensory cells occur simultaneously, which suggests that cells of type I regenerate periodically (Barets and Szabo 1964). This renewal is also observed in other receptor cells of the lateral-line organs (Roth and Szabo 1969). Receptor cells of type II are similar to those of tuberous organs in gymnotids. They are elongated and their basal membrane is attached to the subsensory platform consisting of supporting cells. A larger part of the membrane surface carrying numerous microvilli is oriented into the sensory chamber. Thus, the basal membrane of these cells comes in contact with the afferent nerve terminals forming chemical synapses, and the supporting cells. Mormyromasts are innervated by several myelinated nerve fibers. All cells of type II are innervated with a single nerve fiber, whereas two or three fibers are connected to sensory cells of type I, but each cell comes in contact with only one fiber.

Gymnarchomasts are divided into two types according to their structure (Szabo 1974). Gymnarchomasts of type I are similar to mormyromasts since they contain two kinds of receptor cells. However, they are innervated with single nerve fibers. Generally, the organ consists of a pair of cells, one of each type, but sometimes there may be more than one pair (Szabo 1965). As compared to the common tuberous organs, only a small amount of membrane is in contact with the interior of the lumen. Receptor cells of type I are approximately 1.5 times greater in diameter than cells of type II. The peculiarity of both types of cells is that their apical surface is invaginated and covered with numerous microvilli, the cell nucleus is basally located, and the mitochondria are found in the apical cytoplasm. When approaching the organ, the afferent nerve fiber loses its myelin sheath before penetrating the basement membrane and forms a large axonal enlargement from which fine terminal branches come in contact with the type I receptor cells. Cells of type II are innervated with finer preterminal fibers than cells of type I. The structure of synaptic contacts is similar to that of other electroreceptor cells.

Gymnarchomasts of type II are a more complex anatomical entity composed of 12–13 isolated units innervated by a single nerve fiber. In a separate unit there are one receptor cell and several supporting cells surrounding it (Derbin et al. 1969). The apical membrane is strongly invaginated and possesses a set of microvilli 1.5–4 μm long (Mullinger 1969). The length of the microvilli increases as they approach the base of the invagination. The structure of synaptic contacts does not differ from that of other tuberous organs, and possesses only one type of synaptic contact.

Thus, the data presented testify that the fundamental morphology of the receptor structures of the acousticolateral system of vertebrates shows marked similarity from system to system (auditory, vestibular, lateral-line or electroreceptor) and from species to species. Throughout this system the sensory receptors are hair cells that come in synaptic contact with the afferent nerve fibers of sensory neurons. This makes it possible to attribute mechano- and electroreceptors of the acousticolateral system to the category of secondary sensory structures.

Chapter 2

Electrophysiological Analysis of Synaptic Transmission in Hair Cells

2.1 Electrical Characteristics of Hair Cells. Receptor Potentials

In recent years, considerable progress has been achieved in the study of the electrical properties of hair cells of the acousticolateral system using a variety of physiological techniques including intracellular recording with high-resistance microelectrodes and both membrane patch- and whole-cell recording with gigohm-seal electrodes (Fig. 15) (Lewis and Hudspeth 1983). Using such approaches, electric characteristics of hair cells and their primary responses to adequate stimulation have been thoroughly investigated (Hudspeth 1983, 1985, 1989, 1992; Jacobs and Hudspeth 1990; Pickles 1985; Ashmore 1988a, 1990, 1991, 1992; Roberts et al. 1988; Howard et al. 1988; Ohmori 1989; Cros 1990; Dallos 1992; Dallos and Cheatham 1992; Pickles and Corey 1992).

Since the hair cells of the acousticolateral system are considered to be secondary receptors characterized by the presence of synaptic transmission, it is no less important to analyze presynaptic mechanisms of transmitter release and postsynaptic responses. In this regard, the hair cell is an epithelial mechanoreceptor transducer cell capable of synthesis and release of neurotransmitter.

In 1970 Harris and coworkers were the first to obtain intracellular recordings from the lateral line hair cells of the mudpuppy *Necturus maculosus*. Subsequent intracellular recordings demonstrated the membrane potential of lateral line organ hair cells to be -10 to -65 mV, the input resistance 10–100 mOm, and the membrane time constant 0.2 ms (Flock 1971; Flock et al. 1973a,b). Displacement of the sensory hairs toward the kinocilium evoked a depolarizing receptor potential (RP) up to 5 mV [Fig. 16(5)] (Flock 1971). Displacement in the opposite direction produced hyperpolarizing RP. Thus an important feature of the hair cell's responses to mechanical stimuli is its directionality. It is interesting to note that a 15-mV shift in the membrane potential caused no changes in the conductance of hair cells, i.e., these cells proved to be electrically inexcitable. Moreover, the displacement-response relationship in the dissociated vestibular hair cell of the chick measured at -50 mV and at $+38$ mV were superimposable (Ohmori 1987). Thus, no voltage dependence was observed in gating of mechanoelectrical transduction conductance (Roberts et al. 1990).

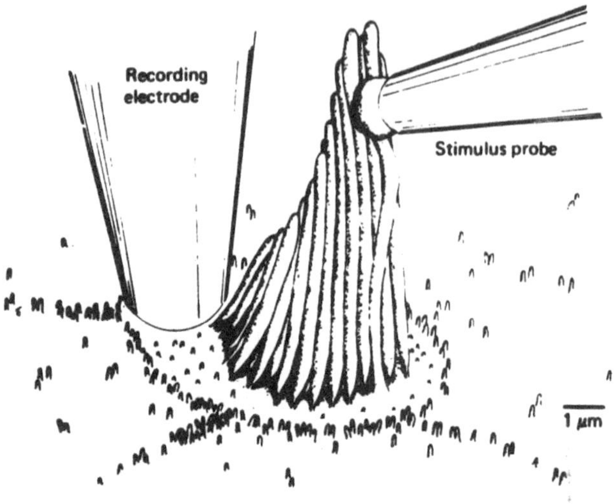

Fig. 15. Diagram of electrical potentials recording in the hair cells. Schematic diagram of the apical surface of one saccular hair cell with a recording electrode and stimulus probe in place. The recording electrode contacts the cell near the hair bundle, where it forms a gigohm seal. Calibrated displacements of the hair bundle are made with the glass stimulus probe moved by a piezoelectric-bimorph element. The stimulus probe is coupled by slight, constant suction to the kinociliary bulb of the hair bundle, which is approximately 7 μm above the surface of the cell. (Holton and Hudspeth 1986a)

Russell and Sellick (1978) provided intracellular recording from mammalian inner hair cells of guinea-pig cochlea. The resting potential of hair cells was shown to be 25–45 mV. In response to pure tone burst, hair cells responded with AC and DC depolarizing RP of the order of 5–17 mV [Fig. 16(2)].

Principally, similar values of the membrane potential were found in the hair cells of the alligator auditory organ (−70 mV), and they responded with a 3-mV potential change to acoustic stimulation (Mulroy et al. 1974; Baden-Kristensen and Weiss 1983; Holton and Weiss 1983; Fuchs and Evans 1988) [Fig. 16(1)]. The hair cells of the macula sacculi in mudpuppy had a resting membrane potential of −70 mV which is sensitive to external K^+ concentration (Bracho and Budelli 1978). The resting potentials of the isolated hair cells of the semicircular canals of the pigeon (Correia et al. 1989) and frog (Housley et al. 1989) were found to be −54 mV and −45 mV, respectively. In goldfish sacculus, two different types of hair cells were described (Sugihara and Furukawa 1989). The short hair cells had a resting potential of about −75 mV, the long ones −90 mV.

Investigations on turtle hair cells established that the membrane potential and RP were found to be −40 to −50 mV and 30–40 mV, respectively and receptor current up to 200 pA (Crawford and Fettiplace 1978; Art and Fettiplace 1987) [Fig. 16(4); 17]. Intracellular recording from the inner hair

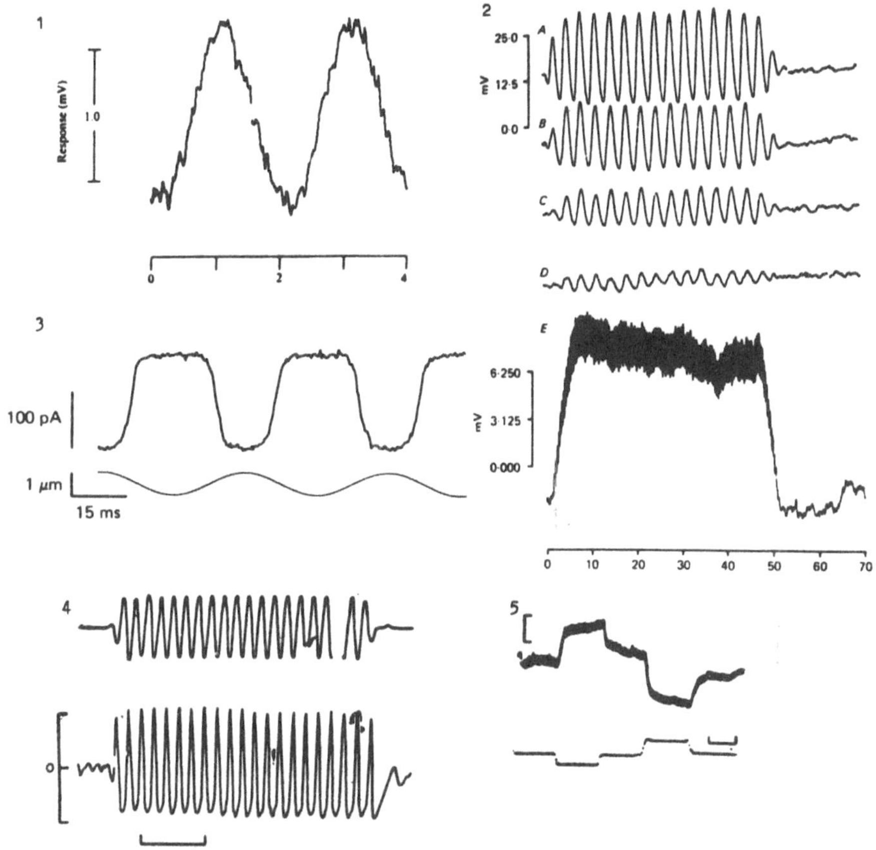

Fig. 16,1–5. Responses of hair cells of the acousticolateral system to mechanical stimulation. **1** Averaged intracellular responses to a continuous tone recorded from a hair cell of the cochlea of the alligator. Time scale, ms. (Mulroy et al. 1974. Reprinted by permission from Nature, vol. 249, N 5456, pp. 493. Copyright 1974. Macmillan Journals Limited). **2** Intracellular receptor potentials from a hair cell of guinea-pig cochlea. *A–D* Responses to 300 Hz tone 90–95 dB in 10 dB steps. *E* Responses to 3 kHz at 80 dB. Time scale, ms. (Russell and Sellick 1981). **3** Four cycles of the receptor current (*upper trace*) evoked by displacing the hair bundle with a 25-Hz sinusoidal stimulus (*lower trace*). Inward (negative) receptor current is plotted *downward*. The stimulus trace is the electrical signal to the probe, with phase adjusted to minimize the hysteresis. A ± 0.5-µm probe displacement at the level of the kinociliary ball corresponds, on average, to a ±71 mrad angular displacement of the stereocilia. (Holton and Hudspeth 1986a). **4** Receptor potentials (*lower trace*) of the cochlear hair cell in the terrapin in responses to mechanical stimulation (*upper trace*). *Horizontal bar* 20 ms; *vertical bar* 20 mV. (Fettiplace and Crawford 1978). **5** Receptor potential (*upper trace*) of hair cell of lateral line of *Necturus maculosus* in response to cupula displacement in the opposite directions (*lower trace*). *Vertical bar* 5 mV; *Horizontal bar* 5 s (Flock 1971)

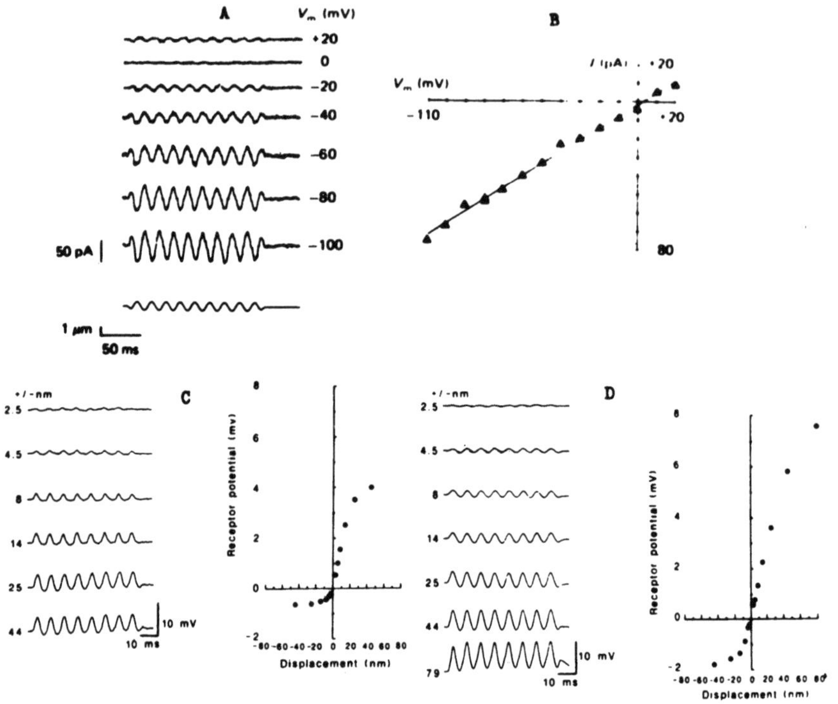

Fig. 17A–D. Current-voltage relation of receptor current in response to gated sinusoidal stimulation of the hair bundle. Receptor-current wave forms measured at different values of membrane potential, V_m, in response to stimulation of the hair bundle by a gated sinusoid of 0.5 µm peak amplitude (*lower trace*). The stimulus frequency was about 39 Hz and the stimulus duration was 256 ms. Wave forms have been digitally smoothed. V_m has been corrected for a Donnan-equilibrium potential of about −7 mV, a value measured in this cell from the difference in reversal potential of current-voltage curves acquired 15 min apart. **B** Current-voltage curve of the same cell. V_m was increased from −110 to +20 mV in 10 mV steps, followed by a return to −80 mV. At each value of V_m, the receptor current was measured in response to gated sinusoid and the amplitude of the steady-state portion of the response was computed from the average of several response cycles. The continuous line, fitted to the data by minimum mean square error criteria, indicates a reversal potential of +3 mV. (Holton and Hudspeth 1986a). **C,D**. Intracellular receptor potentials recorded from an outer and an inner hair cell from the apical low-frequency turn of the mouse cochlea in vitro in response to sinusoidal displacement of their stereocilia. The peak amplitudes of the displacements are indicated *alongside each trace*. The relationship between peak displacements of the stereocilia and receptor potential from the outer and inner hair cells, respectively. (Russell et al. 1986a. Reprinted by permission from Nature, vol. 321, N 6069, pp. 518. Copyright 1986. Macmillan Journals Limited)

cells of the guinea-pig cochlea showed the membrane potential to be −20 to −40 mV, RP being −5 to −17 mV, and cell resistance to be 45 to 60 mOm (Russell and Sellick 1978, 1981, 1983) [Fig. 16(2)]. Later on, Russell and Kössl (1991) also confirmed the earlier findings, that the mean resting potential of inner hair cells was −39 mV. It should be noted that the amplitude of the hyperpolarizing RP is twice as small as that of the depolarizing ones, which is a characteristic feature of electric responses from the hair cells in the acousticolateral system (Hudspeth and Corey 1977; Crawford and Fettiplace 1980; Ohmori 1987).

It is noteworthy that the inner and the outer hair cells differ in their electrical characteristics. For example, in the chick cochlea the inner and outer hair cells had the resting potential −61 and −77 mV, respectively (Fuchs et al. 1988). The resting membrane potentials of mouse inner and outer hair cells in vitro were −30 to −70 mV, and −40 to −70 mV respectively (Russell et al. 1986a,b). A comparative analysis of the electrical characteristics of the inner and outer hair cells in mammals showed that the median membrane potential of the inner hair cells was −20 mV, the depolarizing RP being −25 mV, while that of the outer hair cells was −71 mV, and −15 mV, respectively (Dallos et al. 1982; Dallos 1985a,b, 1986). The input resistance of both cell types ranged from 10 to 34 Ω.

It has been recently demonstrated that the resting membrane potential of mouse inner and outer hair cells in vitro were −30 to −70 and −40 to −70 mV, respectively (Russell et al. 1986a,b). Membrane potential of the isolated outer hair cells from the guinea pig cochlea was −47 mV (Sunose et al. 1992).

Additionally, the RPs of inner hair cells differ in some respects from those of outer ones. For example, depolarizing RPs of the outer hair cells

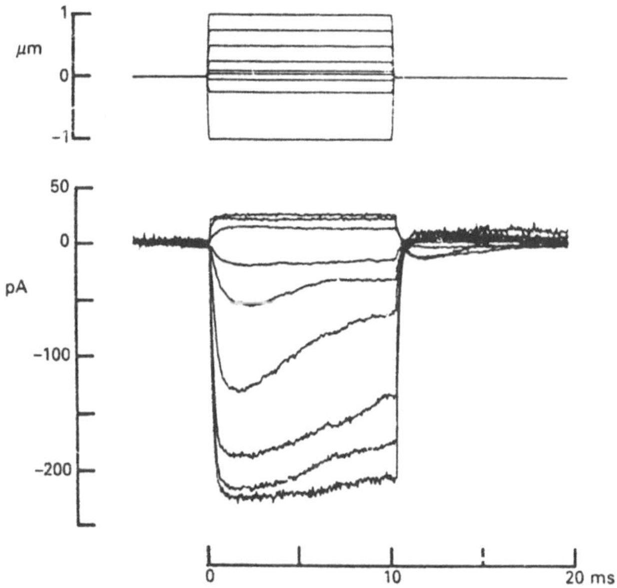

Fig. 18. Averaged transducer currents in an isolated hair cell for displacements of the tip of the hair bundle towards or away from the kinocilium. In the stimulus monitor shown at the top, positive steps indicate movements towards the kinocilium. Inward currents are given as negative relative to the current level in the absence of bundle stimulation at the holding potential, −74 mV. About fifty responses have been averaged for the small traces, and two to ten for large ones. Bundle height, 6.2 μm; resting potential of hair cell, −54 mV; temperature, 23 °C (Crawford et al. 1989)

are saturated at a lower voltage (6.5 mV) than the inner hair cells (9.4 mV) (Fig. 18C,D). The inner hair cells are most sensitive to stereocilia displacement within the range of 0–60 nm, and the outer cells within 0–30 nm. The most sensitive hair cells are depolarized by 0.8 mV when their stereocilia are displaced by 2 nm (Russell et al. 1986a).

Dallos (1985a) believed that inner hair cells are more sensitive than outer ones. However, it was shown that the inner hair cell transducer conductance changed by 0.54 nS when the stereocilia bundle was displaced by 25 nm (21.6 pS per nm) and for the outer hair cells the corresponding value was 0.83 nS by the 8 nm displacement of stereocilia bundle (104 pS per nm) (Russell and Richardson 1987). The outer hair cell input conductance of guinea pig cochlea in vitro ranged form 25 to 40 nS, and typical resting potentials were −15 to −40 mV (Ashmore and Meech 1986).

As mentioned in Chapter 1, the bulk (85–95%) of afferent fibers in the mammal cochlea contact synaptically with the inner hair cells. On the other hand, the outer hair cells represent more than 3/4 of the cochlear receptor cells and receive a large number of efferent endings. The inner hair cells, which represent less than 1/4 of the cochlear receptor cells, receive 90–95% of the afferent innervation. Each inner hair cell is innervated by about 20 neurons. Only 5–10% of the afferent neurons are connected to the outer hair cells, which represent more than 3/4 of the cochlear receptor cells and receive abundant efferent innervation. The information transfer to the central nervous system occurs essentially in the inner hair cell system. The outer hair cells as active mechanical elements influence cochlear micromechanics (Brownell et al. 1985; Crawford and Fettiplace 1986). The outer hair cell function is under efferent control. It has recently been demonstrated that electrical stimulation of efferent nerves markedly reduced the inner hair cell RP of the guinea pig cochlea while causing little change in the resting membrane potential (Brown et al. 1983). In response to electric stimulation of efferent fibers, a hyperpolarizing postsynaptic potential within the range of 8 to 24 mV was observed (Art et al. 1982). During the efferent hyperpolarization there was a reduction of up to 25-fold in the amplitude of depolarizing RP.

An important contribution to the current concepts of the primary process in hair cells was made by investigations on the frog sacculus receptors (Hudspeth and Corey 1977; Corey and Hudspeth 1979a,b; Hudspeth 1983; Eatock et al. 1987). The membrane potential of these cells was −50 to −60 mV and the input resistance 20 Ω; in response to mechanical stimulation a depolarizing RP of 5–10 mV and a receptor current equal to 50–150 nA were generated [Fig. 16(3)]. In the absence of a sensory stimulus, the inward receptor current was 10–20% of its maximal leval. The sign of the receptor potential was determined by the direction of the mechanical stimulus: the displacement of stereocilia towards a kinocilium gave rise to a depolarizing RP and caused a decrease in the input resistance, whereas the stimulation in the opposite direction induced a hyperpolarizing

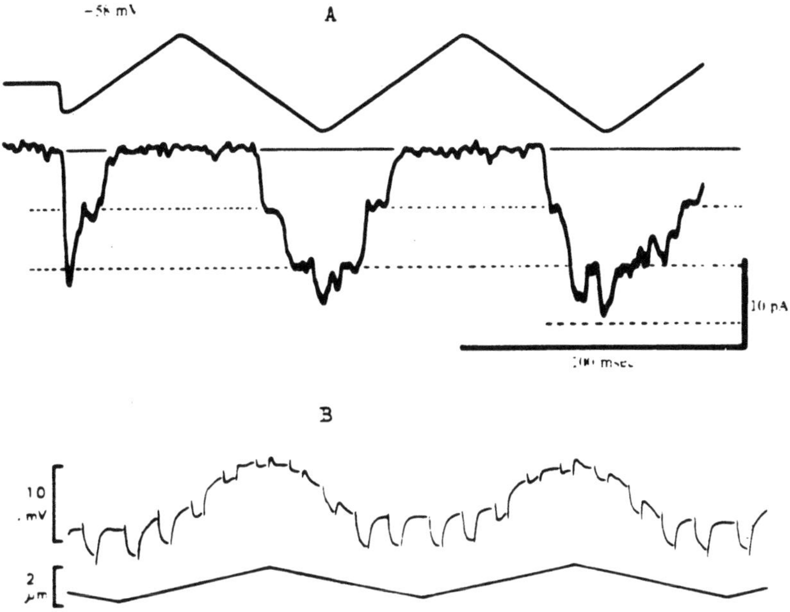

Fig. 19. **A** Transduction current showing discrete levels of current (*lower trace*). Discrete current levels (*dashed lines*) were generated at −58 mV in response to a mechanical stimulation (*upper trace*) of the chick vestibular hair cell. (Ohmori 1984). **B** Changes in the input resistance of a hair cell in the sacculus from bullfrog during its response to mechanical stimulation. Hyperpolarizing square current pulses (70 pA, 45 ms) were injected into a cell while its hair bundle was deflected at 1 Hz (*lower trace*). (Hudspeth and Corey 1977)

RP and an increase in input resistance. The maximal change in input resistance was 50 Ω (Fig. 19B). The threshold sensitivity of the cells was 500 pm, and the latency of the response to rapid stimuli at 22 °C was 40 μs (Corey and Hudspeth 1979b). Within the displacement range up to 0.5 μm, the RP amplitude was linearly dependent on the intensity of mechanical stimuli (Fig. 18B).

It is important to note that even after the removal of the kinocilium (Hudspeth and Jacobs 1979), stereocilia still generated RPs with an amplitude up to 13 mV in response to mechanical stimuli. As in intact cells, the displacement of stereocilia in the direction of the kinocilium still induced a depolarizing RP. In normal cells, a mechanical stimulation of a stereocilium bundle alone caused RP generation, and the stimulation of ten stereocilia provided a receptor potential of 2 mV. Response magnitudes near the auditory threshold for hair cells were quite small, e.g., the most sensitive hair cells were depolarized by 0.8 mV when the tips of stereocilia bundles were displaced by 2 nm (Russell et al. 1986a,b). It was estimated that, at the perceptual threshold of hearing, the bundles of inner hair cells are deflected by 0.1–1 nm. At this deflection a hair cell responds with depolarizing RP of 100 μV or receptor current of 1 pA (Roberts et al. 1988).

Thus the results of the experiments testify to the fact that kinocilia are not actively involved in the transduction processes in vertebrate hair cells. It can be assumed that these structures are essential in providing the polarization of hair cells in ontogenesis and that their role in transduction processes may be confined to a simple transmission of mechanical displacement from the cupula or the otolithic membrane toward stereocilia. The receptor cells of the organ of Corti in adult mammals, birds, and reptiles, are known to contain no kinocilia at all. This probably accounts for the improved receptor responses to high-frequency acoustical stimuli. A fundamental difference should be noted to exist between the hair cells of vertebrates and invertebrates. In the latter, the apical surface has only a kinocilium which transforms mechanical stimuli. In vertebrates, on the contrary, these processes seem to be associated with stereocilia.

Interesting data were obtained while recording extracellular potentials along hair cell stereocilia (Hudspeth 1982). It has been discovered that the

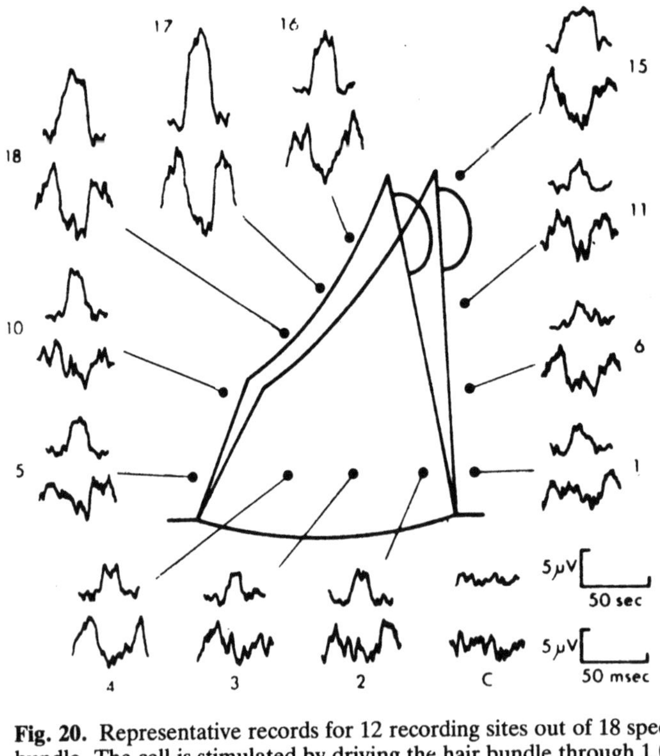

Fig. 20. Representative records for 12 recording sites out of 18 specific sites around a hair bundle. The cell is stimulated by driving the hair bundle through 1.0 μm, 25 Hz, sinusoidal deflections; the hair bundle is diagrammed at rest and a deflection to +1 μm. The *upper record* at each recording site is the output of the lock-in analyzer. The *lower records* are computer-averaged extracellular responses from 5-s samples of electrode output. The control records (*C*) show the noise level in each type of trace. (Hudspeth, 1982)

transduction channels are located at or near the tips of the stereocilia (Fig. 20). The receptor potential recorded from a hair cell was shown to account for 90% of the potential emerging in the stereocilium terminals. On the other hand, mechanoelectrical transduction channels seem to be located at the insertion of the hair bundle into the cuticular plate (Flock 1971; Ohmori 1988a).

More recently Jaramillo and Hudspeth (1991) in the experiments by focally blocking receptor currents with iontophoretically ejected gentamicin showed that the drug was most effective when applied at the top of hair bundle.

Electrophysiological experiments provide strong evidence that transduction channels appear to be gated directly by mechanical stimuli, and involvement of enzymes or second messengers were excluded. These channels have discrete levels of conductance (Fig. 19A). The gating-spring model for transduction has been recently suggested (Fig. 21) (Roberts et al. 1988). The peak transduction conductance of these channels in hair cells is 2–5 nS with the conductance of a single channel being 20–50 pS. Considering the fact that a single hair cell of the vestibular apparatus and the organ of Corti has 40–150 stereocilia, the peak transduction conductance of 2–5 nS could be due to simultaneous opening of a few hundred channels or several channels per stereocilia with conductance of 17 pS (Hudspeth 1982; Russell 1983; Holton and Hudspeth 1986a; Ohmori 1987). Of great interest is the fact that part of the transduction channels are open at rest, and their conductance is 0.14 of the maximal conductance, which would lead to a steady-state inward current of 16 pA at −62 mV (Ohmori 1987). Later on, Russell and Kössl (1991) also proposed that part of the outer hair cell

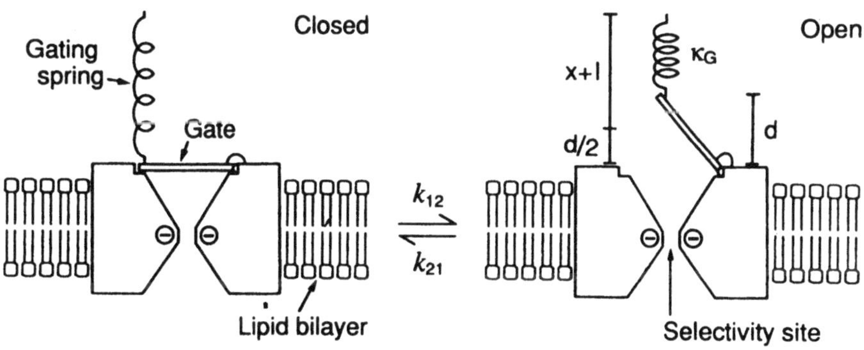

Fig. 21. A model of the transduction element of a hair cell. The cation-selective channel, whose permeability is determined by a negatively charged selectivity site, is regulated by a molecular gate. Positive displacement of the hair bundle increases the tension in the spring attached to the gate. The gate can assume only two positions. When the channel is closed (*left*), the spring is extended by a distance $x + d/r$ beyond its natural length, l. The opening of the channel (*right*) shortens the spring by a distance d. (Howard et al. 1988. Reproduced, with permission, from the Annual Review of Biophysics and Biophysical Chemistry vol. 17, C 1988 by Annual Reviews Inc.)

transducer conductance open at rest and negative current increases and positive decreases of this conductance. The transducer channels of hair cells of the cultured neonatal mouse cochlea of turtle have at least two closed states and one open state (Crawford et al. 1991; Kros et al. 1992).

Two types of stretch-activated ion channels are also present along the lateral wall in guinea pig outer hair cells (Ding et al. 1991). The channels had a conductance of 38–50 pS and 150 pS respectively. These channels could affect the motile responses of the cells by altering the membrane potential or intracellular free Ca^{2+}.

The analysis of current-voltage characteristics has shown that the hair cells behave linearly within a wide range of membrane potential (Fig. 22). For example, the inner and outer hair cells of the guinea pig's organ of Corti were linear over a range of +40 to −150 mV with the membrane potential and the receptor potential being −30 to −55 and −15 to −35 mV, respectively (Fig. 17C,D) (Russell 1983). The reversal potentials for the hair cell receptor current vary considerably in different preparations. They range within −9 to +11 mV in the hair cells of the frog vestibular apparatus (Corey and Hudspeth 1979a,b; Holton and Hudspeth 1986a) (Fig. 18A) and in isolated vestibular hair cells of the chick (Ohmori 1985, 1987); they are close to zero in the guinea pig's vestibular apparatus (Crawford and Fettiplace 1979) and in the mouse cochlear hair cells in organotipic cultures (Russell and Richardson 1987) or correspond to the values of the endocochlear potential of the inner hair cells in the guinea pig's organ of Corti (Russell 1983; Russell and Kössl 1991). Hyperpolarization of hair cells results in a nonlinearity in the current-voltage characteristics of the receptor. The nonlinearity seems to appear due to anomalous and delayed rectification. In the hair cells of the frog vestibular apparatus, anomalous rectification develops at the membrane potential below −100 mV, whereas delayed rectification occurs at a membrane potential of −50 mV (Fig. 22A). The presence of anomalous and delayed rectification determines the relatively small values of hair cell RPs in vertebrates (Corey and Hudspeth 1979a,b). For example, with the membrane potential of −60 mV, the receptor current generated RP of 30 mV, while at the membrane potential of −40 mV the receptor potential was only 0.5 mV. It may be assumed that such peculiarities of hair cells make it possible for them to respond to acoustic stimuli within a wide intensity range.

On the other hand, the relationship between the receptor current and the membrane potential, as well as between the receptor current and the magnitude of hair displacement, is an approximately linear transfer function (Hudspeth and Corey 1977; Russell and Sellick 1983; Holton and Hudspeth 1986a; Ohmori 1987) (Figs. 17B, 22B). For instance, the receptor current in chick isolated hair cells retains linearity during hair displacement up to 0.4 µm. It is essential that, in contrast to electrically excitable and also ACh-activated channels, the activation of the mechanosensitive channels in hair cells is not voltage-dependent (Ohmori 1987).

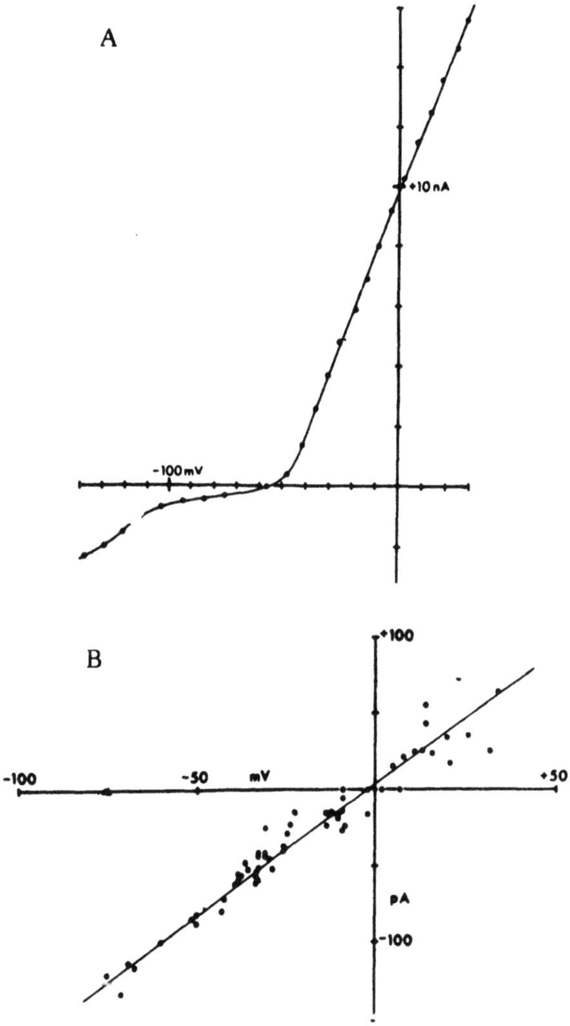

Fig. 22A,B. Current-voltage characteristics for saccular hair cells of bullfrog in the absence of (**A**) and under mechanical stimulation (**B**). (Corey and Hudspeth 1979a. Reprinted by permission from Nature, vol. 281, N 5730, pp. 675, 676. Copyright 1979. Macmillan Journals Limited)

Important characteristics were observed in enzymatically isolated hair cells (Lewis and Hudspeth 1983; Ohmori 1984a; Holton and Hudspeth 1986a; Art and Fettiplace 1987; Russell et al. 1986a). Mechanical stimulation of the frog sacculus hair cells at the membrane potential of a $-60\,mV$ with the latency of $15\,\mu s$ induced a receptor current of $20\,pA$ which reached its maximal value during $100\,\mu s$. A more detailed analysis showed that mechanoelectrical transducer channels have discrete levels of conductances with an elementary conductance of at most 50 pS (Fig. 19A) (Ohmori 1984b). Solitary hair cells of the turtle basilar papilla had resting potential of about −50 mV and maximal transducer current up to 200 pA (Art and Fettiplace 1987).

Recently, detailed analyses of adaptation in hair cells of the bullfrog sacculus have been undertaken (Eatock et al. 1987; Hacohen et al. 1987; Assad et al. 1989, 1991; Crawford et al. 1989; Assad and Corey 1992; Corey and Assad 1992; Fettiplace 1992; Kimitsuki and Ohmori 1992). The saccular neurons adapted fully within 10–50 ms. The microphonic potential of the hair cells adapted within a similar time course. The authors believe that adaptation took place in the hair cell or its mechanical inputs due to a shift in the displacement-response relationship along the displacement axis in the direction of the adapting step. The displacement shift occurred in response to both the depolarizing and hyperpolarizing stimuli. The process is characterized by high temperature, Ca^{2+}, and voltage dependence. It has been recognized that Ca^{2+} is required for hair-cell adaptation (Assad et al. 1989; Crawford et al. 1991; Kimitsuki and Ohmori 1992). Kimitsuki and Ohmori (1992) assumed that Ca^{2+} entry through the transduction channels must pass through a Ca^{2+} buffer site before reaching the adaptation site of the transducer channel. It is interesting that a high extracellular Ca^{2+} concentration increases the rate of adaptation. Ca^{2+} ions enter the stereocilia through the transduction channels and their site is very near these channels (Assad et al. 1989). The process of adaptation was modified by pharmacological blockers of calmodulin (Corey et al. 1987). The shift process seems to occur at a stage preceding the gating of transduction channels and may be due to a gradual reduction in the tension in the gating springs (Howard and Hudspeth 1987). The most likely structures to be identified as the gating springs are the "tip links" connected adjacent stereocilia (Pickles et al., 1984). Assad et al., (1991) proposed that the tip links convey tension to the transduction channels and responsible for adaptation of the transducer current. Indeed, the low – Ca^{2+} solution irriversibly eliminated the tip links and the transducer current. Lately, Assad and Corey (1992) put forward the active motor model for adaptation in hair cells.

2.2 Ionic Mechanisms of Receptor Potential Generation

The apical membrane of the hair cells in the acousticolateral system is known to be bathed in medium with an elevated K^+ concentration, which is thought to be important for the primary processes in receptors (Ilyinsky 1975). Some authors, however, assume that the asymmetrical distribution of K^+ is not essential for the transduction processes (Ohmori 1984a). Since the dominant cation in the endolymph of the inner ear is K^+, this ion probably carries the major part of the receptor current (Matsuura et al. 1971; Corey

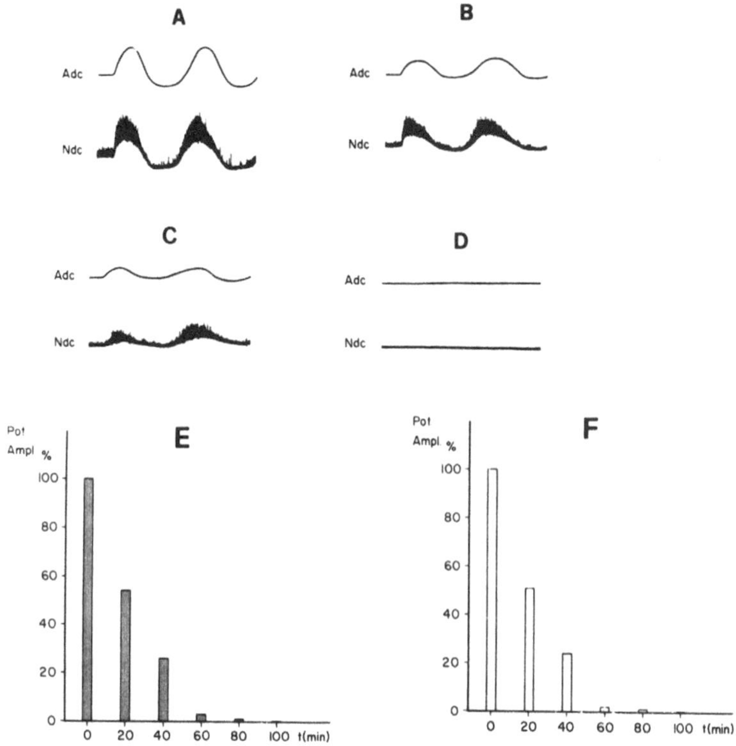

Fig. 23A–F. Effects of endolymphatic and perilymphatic K^--free solutions on slow ampular (*Adc*) and nerve (*Ndc*) potentials. Recordings: 10 min (**A**); 20 min (**B**); 40 min (**C**); 80 min (**D**) after exposure to K^+-free solutions. Histograms: the amplitude of the ampular (**E**) and nerve (**F**) potentials (mean values from five experiments expressed as percentages of controls). (Valli et al. Acta Oto-Laryngol (Stockh) 1977, v. 84: 347)

and Hudspeth 1979a; Russell 1983). The involvement of other ions in RP generation has not been completely elucidated so far. Thus, in goldfish sacculus, a complete substitution of Na^+ for K^+ in the endolymph did not affect the microphonic potential (Matsuura et al. 1971). In contrast, it was shown that in the absence of K^+ in the endolymph, the impulse activity in the ampullary nerve fibers of frog semicircular canals was blocked irrespective of its ionic content (Valli et al. 1977, 1979; Zucca et al. 1982) (Fig. 23). Interestingly, like Cs^+, Rb^+ was able to substitute K^+ in maintaining receptor activity (Valli et al. 1984).

The mechanoelectrical transduction channels are nonspecific by nature and a wide variety of monovalent, divalent, and organic cations with diameters up to 0.7 nm are permeable through these channels in the hair cells of the goldfish inner ear, the lateral-line organs of the mudpuppy, the bullfrog sacculus, and the vestibular organ of chicks (Matsuura et al. 1971; Sand 1975; Corey and Hudspeth 1979a; Ohmori 1985). The relative amplitudes of the receptor current for hair cells of bullfrog sacculus were: Li^+, 0.9; Na^+, 0.9; K^+, 1.0; Rb^+, 1.0; Cs^+, 1.0; NH_4, 1.3 (Corey and Hudspeth 1979a). The transduction channels are also permeable to tetraethylammonium (TEA) with molecules having a diameter of 0.7 nm in an unhydrated state. Hence these channels have a diameter of at least this dimension and resemble the nonspecific ACh-activated channels of the motor end-plate (Maeno et al. 1977; Dwyer et al. 1980), and the mechanosensitive channels of the Pacinian corpuscle (Akoev et al. 1980c). Therefore, the receptor current is curried in vitro by whatever ions are available. In vivo this would normally be K^+.

Calcium ions have long been known to be essential for the mechanoelectrical transducer process, since their removal from the solution or the addition of Co^{2+}, La^{2+}, or Mg^{2+} blocked the responses from the hair cells of the lateral-line organs in the mudpuppy (Sand 1975), frog (Jørgensen 1983, 1984, 1988), and fish (Baumann and Roth 1986), the toad sacculus (Corey and Hudspeth 1979a), in the frog semicircular canals (Valli et al. 1979; Zucca et al. 1982), and vertebrate cochlea (Tanaka et al. 1980; Fettiplace 1992). Corey and Hudspeth (1979a) were able to demonstrate that the transducer current in the sacculus of the bullfrog in isotonic Ca^{2+} solution without permeant monovalent cations was 0.3 of that produced in solution containing 130 mM K^+. However, considering that Ca^{2+} concentration in the endolymph is equal to 30 μM (Ikeda et al. 1987; Crawford et al. 1991), it may be assumed that these ions can carry very little of the receptor current in vivo. For example, Na^+, Ca^{2+}, and Mg^{2+} together carry about 1% of the total receptor current and the Ca^{2+} current through a single transducer channel seems to be 4–12 fA. Nevertheless, their presence in the solution is indispensable, since the reduction of their concentration to below 10 μM blocks the receptor current completely (Corey and Hudspeth 1979a; Ohmori 1985; Crawford et al. 1991). It is noteworthy that Sr^{2+} can replace Ca^{2+} in this role but Ba^{2+} and Mg^{2+} cannot (Sand 1975; Corey and Hudspeth 1979a; Ohmori 1985). The permeability sequence of the divalent cations through the transducer channels was $Ca^{2+} > Sr^{2+} > Ba^{2+} > Mn^{2+} > Mg^{2+}$ (Ohmori 1985). The transducer channels of the vestibular hair cells of chick (Ohmori 1985) and guinea-pig cochlea (Ashmore and Meech 1986) have much greater permeability to divalent cations than to monovalent ones. Ohmori (1985) suggests that, in contrast to the mechanosensitive channels of the Pacinian corpuscle (Akoev 1982), the channels of hair cells have a site of relatively high negative field strength at the selective filter.

More detailed analysis with the Ca^{2+}-sensitive dye Fura-2 has revealed an increase and a decrease in the fluorescence ratio in the hair bundle with depolarizing and the hyperpolarizing stimulation, probably due to changes in the intracellular Ca^{2+} concentration (Ohmori 1988b). The changes in fluorescence intensity are most likely due to the influx of calcium ions through the mechanoelectrical transduction channels. The highest fluorescence intensity ratio was observed near the hair bundle insertion into the cuticular plate. In the resting state in both outer and inner cochlear hair cells of the guinea pig, the amount of Ca^{2+} just beneath the plasma membrane was greater than that throughout the entire cytoplasm (Yamashita et al. 1990, 1991; Chabbert et al. 1991). High K^+ in the external solution led to a marked increase in Ca^{2+} concentration inside the cells, which was not observed in Ca^{2+}-free solution. It has recently been shown that ACh (Shigemoto and Ohmori 1990) and ATP (Ashmore and Ohmori 1990; Dulon et al. 1991) control the intracellular calcium level in hair cells. A cyclic-AMP-gated conductance of about 16 pS was shown to exsist in the plasma membrane of the guinea-pig cochlea hair cells (Kolesnikov et al. 1991).

A change in the concentration of mono- and divalent ions in the apical membrane of electrosensory hair cells also affects their activity (Roth 1971; Peters et al. 1975; Zhadan and Zhadan 1975; Bauswein 1977). The elevation of the external Ca^{2+} concentration from 0.4 up to 5.0 mM was found to have no influence on the resting activity in spite of the rising threshold of the ampullary electroreceptors of *Ictalurus* (Zhadan and Zhadan 1975). In Ca^{2+}-free solution, the receptors did not respond to electric stimuli. The rise in the K^+ concentration from 5 to 20 mM caused an increase in the resting activity and a decrease in the receptor threshold (Zhadan and Zhadan 1975).

Various lines of evidence suggest that the sensory hair cells are capable of producing some form of motor activity (Brownell et al. 1985; Zenner et al. 1985; Flock et al. 1986; Reuter and Zenner 1990). Furthermore, actin has been shown to be present in both the cuticular plate and the stereocilia, while myosin and α-actin are present only in the cuticular plate (Drenckhahn et al. 1982). It has been recently demonstrated that actin may be important in the functioning of hair cells in the cochlea (Barron et al. 1987). Cytochalasin D, an inhibitor of actin polymerization, reduced the amplitude of the compound action potential (CAP) in the guinea-pig cochlea (Barron et al. 1987). There is growing evidence that outer hair cells influence, possibly through contraction, the microdynamics of the basilar membrane electric activity of inner hair cells, and thereby contribute to the frequency selectivity of the mammalian auditory system (Brown et al. 1983).

Direct experiments have established that hyperpolarization leads to elongation, while depolarization induces a shortening of outer hair cells by several micrometers (Brownell et al. 1985). A marked asymmetry appears to exist in the changes in length during hyperpolarization and depolarization (Santos-Sacchi 1989). For a given stimulus magnitude, elongation of outer hair cells is as much as 50% greater than shortening. At normal resting potentials in vivo, displacements are about 2 nm/mV, but increase up to 15 nm/mV as the cell is depolarized (Santos-Sacchi 1989). While in vivo intracellular recordings have clearly shown that both inner and outer hair cells are capable of generating RP in response to acoustic stimulations (Dallos et al. 1982; Russell and Sellick 1983), it is becoming increasing clear that outer hair cells additionally function as effectors. The electrical activity of outer hair cells is thought to modify the mechanical properties of the same cells and, in turn, to influence organ of Corti micromechanics. Thus, an indirect micromechanical interaction between outer and inner hair cells may account for the exquisite tuning characteristics and sensitivity of the eighth nerve fibers which predominantly (90–95%) innervate the inner hair cells. Highly interesting are the findings of Brownell et al. (1985), who has demonstrated that, in contrast to outer hair cells, mechanical responses were absent in the inner hair cells.

Analysis of the effect of various ions on mechanical responses of outer hair cells has shown that replacement of intracellular K^+ with Na^+ does not affect the movement of these cells (Ashmore 1987). Moreover, inward Ca^{2+} current is not required for eliciting mechanical responses in the outer hair cells (Santos-Sacchi and Dilger 1988). The authors believe that outer hair cell movements appear to depend upon membrane potential rather than on transmembrane currents.

It is noteworthy that inhibitors of ATP synthesis, drugs against actin and tubulin, failed to inhibit the changes in length (Holley and Ashmore 1988). The authors conclude that the motile mechanism is unlike most other forms of cell motility. It has been shown that phosphoinositides mediate motility in cochlea outer hair cells (Schacht and Zenner 1987).

The properties of mechanoelectrical transduction in hair cells have recently been analyzed by means of various approaches. It was shown, for example, that cobalt ions completely block the mechanosensitivity of the lateral line of roach (Karlsen and Sand 1987) due to competitive inhibition of Ca^{2+} component of receptor current (Sand 1975; Baumann and Roth 1986), which changes the permeability of the transducer membrane to monovalent ions. Among the possible effects of Ca^{2+} may be the influence on the stiffness of the stereocilia (Orman and Flock 1983), or the machanical properties of the cytoskeleton (Hudspeth 1983).

Another tool for analyzing transducer processes in hair cells is the use of aminoglycoside antibiotics, that are known to affect hair cell neural activity at a concentration between 1 and 10 mM (Wersäll and Flock 1964; Bernard 1983; Kroese and van den Bercken 1980, 1982; Kroese et al. 1989; Jaramillo and Hudspeth 1991; Denk et al. 1992). The influence of aminoglycoside antibiotics may be due to interference with Ca^{2+}-dependent membrane phenomena, since Ca^{2+} antagonized their effect. Antibiotics are also assumed to affect the mechanical properties of the sensory hairs (Kroese and van den Bercken 1980, 1982; Howard and Hudspeth 1988) and alter the synaptic transmission (Bernard 1983) and mechanoelectrical processes (Kroese and van den Bercken 1982; Bernard 1983; Kroese et al. 1989). Available evidence indicates that aminoglycoside antibiotics reversibly block the receptor current of bull-frog hair cells. Interestingly, drugs affect the receptor current only when they are applied to the fluid bathing the apical surfaces of the hair cells, and interfere with mechanoelectrical transduction via a voltage-dependent blockade of transduction channels (Kroese et al.

1989). On the contrary, intracellular infusion of the drugs does not block transduction current.

In the tight epithelia, Na^+ channels of the apical membranes are reversibly blocked by a low concentration of amiloride (Sariban-Sohraby and Benos 1986). In the lateral line of *Xenopus* (Jørgensen 1984), amiloride reversibly reduced mechanosensitivity. Recently, a more detailed analysis was performed of the effect of amiloride on mechanoelectrical transduction channels in dissociated hair cells of a chick cochlea and vestibular organs (Jørgensen and Ohmori 1988). Amiloride was shown to block these channels' conductance in a dose-dependent manner with a dissociated constant of 5×10^{-5} M. The voltage-dependent Ca^{2+} channels of the basolateral membrane were not affected by amiloride.

Recently, a detailed study of the properties of a set of ionic channels in hair cells was carried out (Lewis and Hudspeth 1983; Ohmori 1984a; Art et al. 1986; Hudspeth 1986; Ashmore and Meech 1986; Ashmore 1988b; Hudspeth and Lewis 1988a; Fuchs and Evans 1988; Santos-Sacchi and Dilger 1988; Lang and Correia 1989; Housley et al. 1989; Sugihara and Furukawa 1989). Hudspeth (1986) believes that there are at least seven ionic conductances in the bullfrog's saccular hair cells. Depolarization of hair cells in voltage-clamp experiments produced several specific inward and outward ionic currents (Fig. 24). The inward Ca^{2+} current (240 pA) is activated at a potential more positive than −60 mV and was maximal at about −10 mV (Fig. 25A). The time constant of its activation at a membrane potential −20 mV and 22 °C is equal to 4 ms. This inward current is carried by Ba^{2+} and also Sr^{2+}, and is blocked by Ni^{2+} and Cd^{2+}. Its main characteristics include its partial activation at the resting potential, and its inactivation is not appreciable during sustained depolarization lasting 200 ms. These characteristics of Ca^{2+} current account for the tonic discharges in afferent fibers due to the release of neural transmitter. This allows the afferent fibers to encode responses to both positive and negative deflections of hair bundles. There are between 200 and 500 Ca^{2+} channels on the basal part of the hair cell membrane (Ohmori 1984a) or about 1600 Ca^{2+} channels in hair cells from frog crista ampullaris (Prigioni et al. 1992). It has been recently demonstrated that L-type Ca^{2+} channels are directly involved in the transmitter release at afferent synapses (Bobbin et al. 1990; Roberts et al. 1990; Prigioni et al. 1992). Nakagawa et al. (1991) also concluded that the Ca^{2+} channels in OHSs of guinea pig cochlea are of the L-type.

It is interesting that in hair cells isolated from the semicircular canals of the frog, Na^+ channels, in addition to the Ca^{2+} channels, seem to participate in generation of the spike potential (Sugihara and Furukawa 1989). This potential was shown to be resistant to tetrodotoxine (TTX). Na^+-channel activity has also been shown in alligator hair cells (Evans and Fuchs 1987), although the presense of Ca^{2+} spikes has been noted in hair cells of alligator cochlea (Fuchs et al. 1988), frog sacculus (Hudspeth and Corey 1977), and ampullae of Lorenzini (Bennett and Obara 1986).

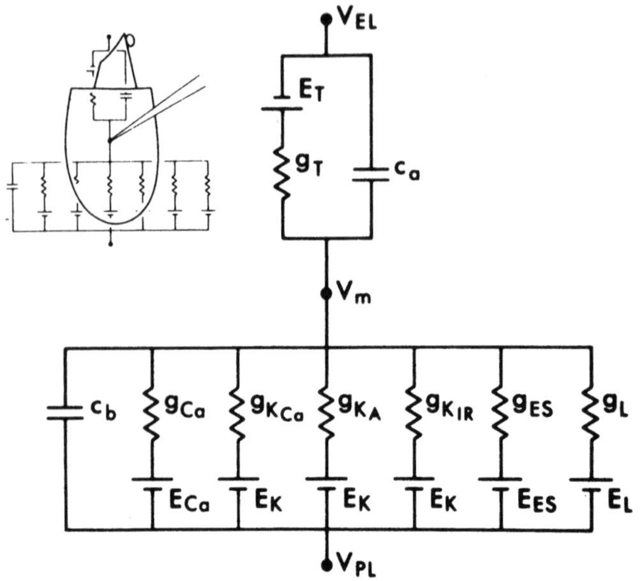

Fig. 24. Schematic diagram of the ionic conductances of the bullfrog's saccular hair cell. The plasmalemma is divisible into apical and basolateral components (*inset*); at least some of the conductances are restricted to one or the other of these regions. Within the hair bundle occurs the transduction conductance, g_T, with a reversal potential, E_T, near 0 mV. The basolateral membrane contains a voltage-sensitive Ca^{2+} conductance, g_{Ca}, with the associated Nernst potential, E_{Ca}. There are three conductances whose Nernst potentials are at E_K: a Ca^{2+}-sensitive K^+ conductance, $g_{K_{Ca}}$, an A conductance, g_{KA}, and an inwardly rectifying conductance, g_{KIR}. The nature of the conductance activated by the efferent synaptic transmitter, g_{ES}, and its reversal potential, E_{ES}, are not known in this cell. The cell leakage conductance, g_L, is associated with a reversal potential, E_L, estimated at +30 mV. The apical and basolateral membrane surfaces also have significant capacitances, respectively C_a and C_b. V_{PL} is defined as zero. (Hudspeth 1986) Reprinted with permission from Hearing Research v. 22, pp. 21 Copyright 1986, Elsevier Scientific Publishers)

Besides the Ca^{2+}-current, depolarization of hair cells also evokes four potassium currents, namely, transitory A-type current, Ca^{2+}-activated K^+ conductance, an inwardly rectifying current and voltage-gated current (Fig. 25B,C) (Lewis and Hudspeth 1983; Hudspeth and Lewis 1988a; Ohmori 1984a; Ashmore and Meech 1986; Art and Fettiplace 1987; Housley et al. 1989; Sugihara and Furukawa 1989; Kros and Crawford 1990; Fuchs and Evans 1990; Rennie and Ashmore 1991; Steinacker and Romero 1991; Griguer et al. 1993). The main characteristics of A current include its blockade by Cs^{2+}, its insensitiveness to external TEA, and its blockade in a voltage-dependent manner by 4-aminopiridine (4-AP) (Lewis and Hudspeth 1983). It becomes activated at voltages more positive than −50 mV, the time constant of ionic activation and inactivation being 80 ms at −40 mV. In

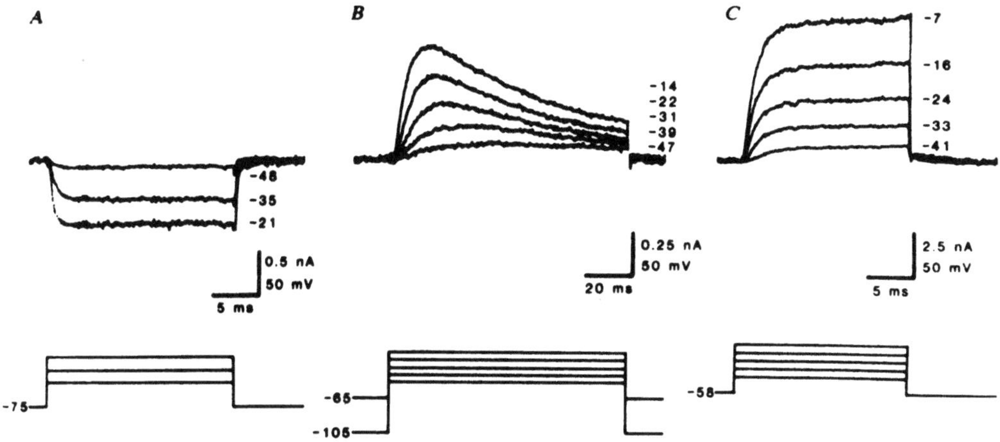

Fig. 25A–C. Voltage-clamp recordings from isolated hair cells showing the three ionic currents. In each panel the *upper set of traces* shows membrane currents and the *lower set* represents command voltage. To the *right* of each current trace, the corresponding command voltage is shown in mV; the holding potential in mV is shown to the *left* of the voltage command traces in each panel. **A** Ca^{2+} current. **B** Transient K^+ current (A current). **C** Ca^{2+}-activated K^+ current. (Lewis and Hudspeth 1983). Reprinted by permission from Nature, vol. 304, N 5962, pp. 539. Copyright 1983. Macmillan Journals Limited)

semicircular canal hair cells from the pigeon there are two types of voltage-dependent calcium-insensitive K^+ conductances (Lang and Correia 1989). The rapid K^+ conductance was inhibited by 4-AP, whereas the slow K^+ conductance was suppressed by TEA. Two types of A-current in isolated hair cells from the crista ampullaris of the frog have been recently described (Norris et al. 1992).

The recent studies have shown the presence of voltage-gated (delayed rectifier) potassium current in hair cells (Fuchs and Evans 1990; Kros and Crawford 1990; Steinacker and Romero 1991). No differences in voltage activation range, pharmacology, or kinetics of the Ca^{2+} current were found in tall hair sells from apical and basal regions of the cochlea of the chick (Fuchs et al. 1990).

Another outward K^+ current is the Ca^{2+}-activated K^+ current. It is activated at a potential above −60 mV with a time constant of 3–5 ms, and is blocked by external TEA ($K_{0.5}$ = 1 mM), Ni^{2+}, and Cd^{2+}, being insensitive to 4-AP. It has been established on isolated hair cells of the turtle auditory organ that a single cell contains about 1200 such potassium channels (Art and Fettiplace 1987). Moreover, outer hair cells of guinea-pig cochlea have been shown to possess a Ca^{2+}-activated K^+ conductance at the base of the cell (Ashmore and Meech 1986). Single-channel recordings revealed two types of Ca^{2+}-activated K^+ channels with conductance of 240 pS and 45 pS. The activity of these channels is determined by Ca^{2+} influx through the

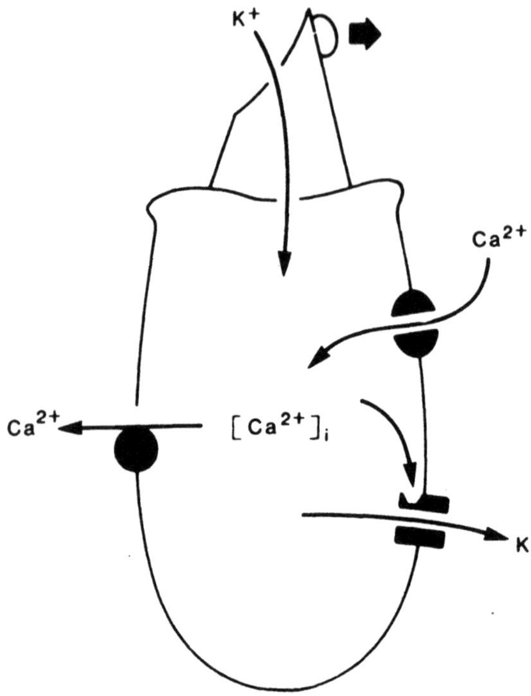

Fig. 26. Schematic diagram of the proposed mechanism for resonant frequency tuning in the bullfrog's hair cell. Depolarization of the cell by a stimulus opens voltage-sensitive Ca^{2+} channels. The *Ca^{2+}* influx depolarizes regeneratively the cell; *$[Ca^{+}]i$* rises, thus activating Ca^{2+}-sensitive K^{+} channels. The resulting *K^{+}* efflux repolarizes the cell to and beyond the steady-state potential. Ca^{2+} within the cytoplasm is buffered and rapidly extruded. ((Hudspeth 1986) Reprinted with permission from Hearing Research v. 22, pp. 25, Copyright 1986, Elsevier Scientific Publishers)

apical transducer channel. It is worth noting that the number of these K^{+} channels in hair cells of the turtle cochlea range from about 480 to 3700 per cell (Art and Fettiplace 1987) and about 700 per frog sacculus hair cell (Roberts et al. 1990). The Ca^{2+} and Ca^{2+}-activated K^{+} channels are arranged in average in of 20 clusters on the basolateral membrane surface. Each channel cluster conteins about 90 Ca^{+} and 40 Ca^{2+}-activated K^{+} channels (Roberts et al. 1990). What may the functional role of these conductances be? On the one hand, Ca^{2+}-activated K^{+} conductance clamps the membrane potential close to K^{+} equilibrium potential and thereby maximizes the inward driving force for cations across the transduction channels of the apical membrane (Ashmore and Meech 1986). On the other hand, Ca^{2+}-activated K^{+} conductance is thought to contribute to the electrical resonance of hair cells (Fig. 26) (Lewis and Hudspeth 1983; Art et al. 1986; Hudspeth 1986; Art and Fettiplace 1987; Hudspeth and Lewis 1988b). The most important factor underlying the resonance frequency is the rate of K^{+}

conductance activation, which varies over two orders of magnitude. The density of K^+ channels in hair cells increases with an increase in the resonance frequency. The question remains to be answered as to how the hair cells regulate the number and kinetics of their K^+ channels (Art and Fettiplace 1987). Several types of Ca^{2+}-activated K^+ channels were also revealed in goldfish hair cell saccular membrane (Sugihara and Furukawa 1896). On the other hand, in the same object, three different type of inwardly rectifying K^+-channels, i.e., inactivating, sustaining, and slow, were described (Furukawa 1986). This current could contribute to setting the resting membrane potential of hair cells (Fuchs and Evans 1990).

2.3 The Functional Properties of Hair Cells of Ampullary Electroreceptors

2.3.1 Electrical Sensitivity of Electroreceptors

The morphological and physiological features of electroreceptors have been reviewed repeatedly (Dijkgraaf 1963; Akoev et al. 1971; Bennett 1971; Szabo 1974; Fessard and Szabo 1974; Murray 1974; Kalmijn 1974; Andrianov et al. 1979; Viancour 1979; Bullock 1982; Andrianov and Ilyinsky 1983; Brown and Ilyinsky 1984; Zakon 1986; Akoev and Andrianov 1989). Recently, ampullary electroreceptors have been found in the cyclostomes (Bodznick and Northcutt 1981; Akoev and Muraveiko 1984), in axolotl (Munz et al. 1982) and platypus (Scheich et al. 1986; Gregory et al. 1988). In this chapter we shall focus on the recent knowledge on ampullary receptors, primarily those of nonteleost fish (skates, rays) and the marine catfish, *Plotosus*. Investigations on these receptors have made a great contribution to our understanding of the primary mechanisms of electroreception.

Murray (1960) was the first to discover the high sensitivity of the ampullae of Lorenzini to a weak electric field. More recently, the high sensitivity of the ampullae of Lorenzini to electric current has been demonstrated in behavioral and electrophysiological experiments (Murray 1962a; Dijkgraaf and Kalmijn 1966; Kalmijn 1971; Bennett 1971; Akoev et al. 1976a; Bromm et al. 1976; Brown and Ilyinsky 1984). It has also been found that the ampullae of Lorenzini respond to mechanical stimuli (Murray 1960; Loewenstein 1960; Akoev et al. 1976a), to changes in salinity (Murray 1962a; Loewenstein and Ishiko 1962; Akoev et al. 1980d), to thermal stimuli (Sand 1938; Hensel 1955; Brown et al. 1985; Akoev et al. 1980d; Brown and Ilyinsky 1984), and to magnetic stimuli (Andrianov et al. 1974; Kalmijn 1974, 1982; Akoev et al. 1976a; Brown and Ilyinsky 1978).

Unlike the receptors of the organ of Corti, the vestibular apparatus, and lateral-line organs, no microelectrode or whole-cell patch-clamp investigations of electrosensory cells have been carried out so far. In many investigations the characteristics of electroreceptors were analyzed by recording

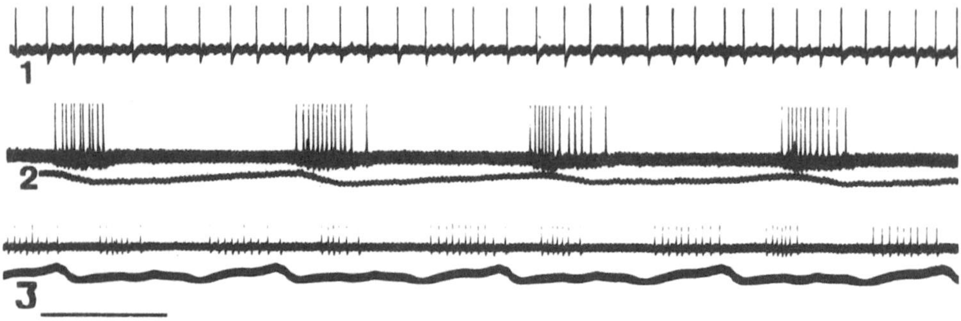

Fig. 27,1–3. Types of background discharge of the electroreceptors of the skate *D. pastinaca*. **1** Fiber with rhythmical activity (artificial respiration of fish). **2,3** Fibers with burst discharges (natural respiration of fish). *Upper trace* fiber discharge; *lower trace* respiration (for **2** and **3**). *Upward movement of beam* corresponds to expiration. Time bar 500 ms. (Akoev et al. 1976a) (Reprinted with permission from Comp. Biochem. Physil. v. 53A Akoev et al. Physiological properties of electroreceptors of marine skates, Copyright 1976, Pergamon Press PLC)

impulse activity from single afferent units. The presence of ampullary canals up to several centimeters long makes it possible to use the current and voltage clamp technique (Clusin and Bennett 1977a,b; Obara 1976; Brown and Ilyinsky 1984). Properties of the ampullae of Lorenzini in the Black Sea skates *Raja clavata* and *Dasyatis pastinaca* were thoroughly analyzed on the receptors in vivo and in vitro (Akoev and Ilyinsky 1972, 1973; Akoev et al. 1974, 1975, 1976a,b, 1980d; Akoev and Andrianov 1989).

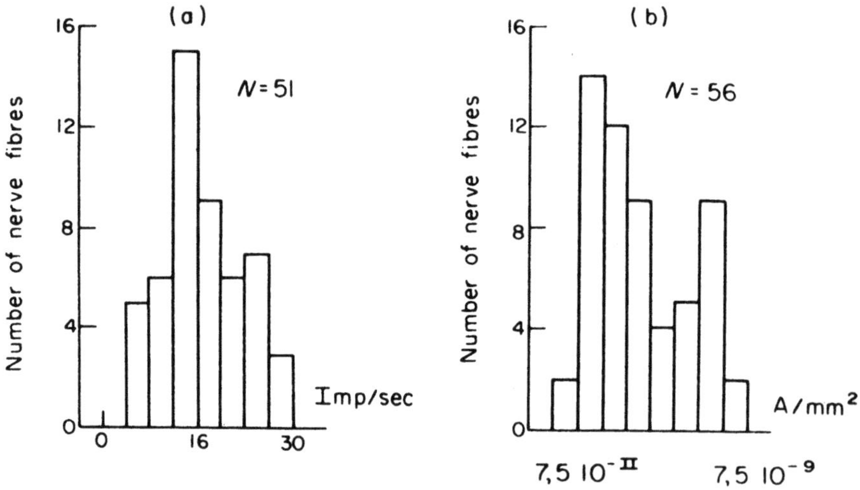

Fig. 28a,b. Distributions of frequency of background activity (**a**) and of thresholds (**b**) of the electroreceptors of the skate *D. pastinaca*. t = 16 °C. (Akoev et al. 1976a) (Reprinted with permission from Comp. Biochem. Physil. v. 53A Akoev et al. Physiological properties of electroreceptors of marine skates, Copyright 1976, Pergamon Press PLC)

It should be emphasized that under natural respiration all receptors exhibited a respiratory rhythm [Fig. 27(2,3)]. Ampullary receptors may be stimulated by local fields caused by respiratory muscle potentials (Dijkgraaf and Kalmijn 1966; Peters and Buwalda 1972; Akoev et al. 1976a; Montgomery 1984). Akoev et al. (1976a) believed that the fish was able to perceive the members of the species and to detect prey through the distortion of their own respiratory field potential. Subsequently, Montgomery (1984) demonstrated that this respiratory burst activity is minimized by a common mode rejection mechanism in the dorsal nucleus of the medulla. On the contrary,

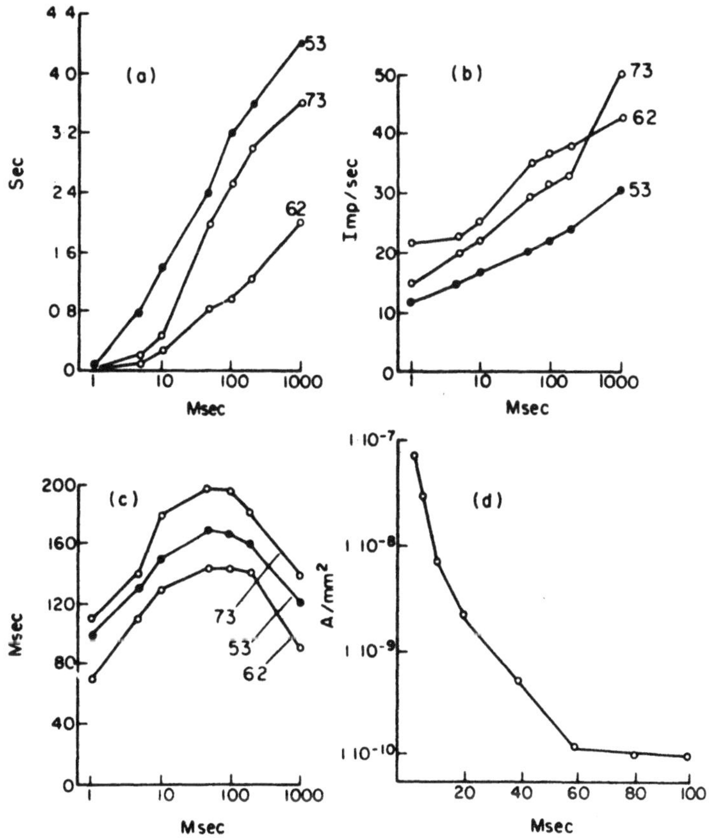

Fig. 29a–d. Effect of duration of cathodic pulses on the responses of the ampullae of Lorenzini of the skate *D. pastinaca*. **a** Relation between the duration of applied pulse (*abscissa*) and duration of inhibition of impulse activity after termination of the pulses. **b** Relation between the frequency of OFF-responses and the duration of pulses (*abscissa*). **c** Relation between the latency of OFF-responses (*ordinate*) and duration of pulses. **d** Relation between the current density threshold and duration of pulses, t = 16 °C. Numbers of neurons at the *right* of each curve. (Akoev et al. 1976a) (Reprinted with permission from Comp. Biochem. Physil. v. 53A Akoev et al. Physiological properties of electroreceptors of marine skates, Copyright 1976, Pergamon Press PLC)

in fish under artificial respiration, all units showed fairly regular discharge rate ranging from 10 to 30 imp/s [Figs. 27.(1), 28a]. For single afferent fibers, thresholds in vitro range from a fraction of a microvolt to a few microvolts (Bromm et al. 1976). The behavoral thresholds may be lower, i.e., 5 nV/cm (Kalmijn 1982). The most sensitive units of skate had a resting rate of 10–20 imp/s and threshold of 10^{-9} to 10^{-11} A/mm^2 (Fig. 28a,b). The threshold of ampullae of Lorenzini increased with a decrease in stimulus duration (Fig. 29d). The shortest stimulus duration when receptor responses could be obtained was found to be 0.3 ms, the threshold being increased by 10^3 times.

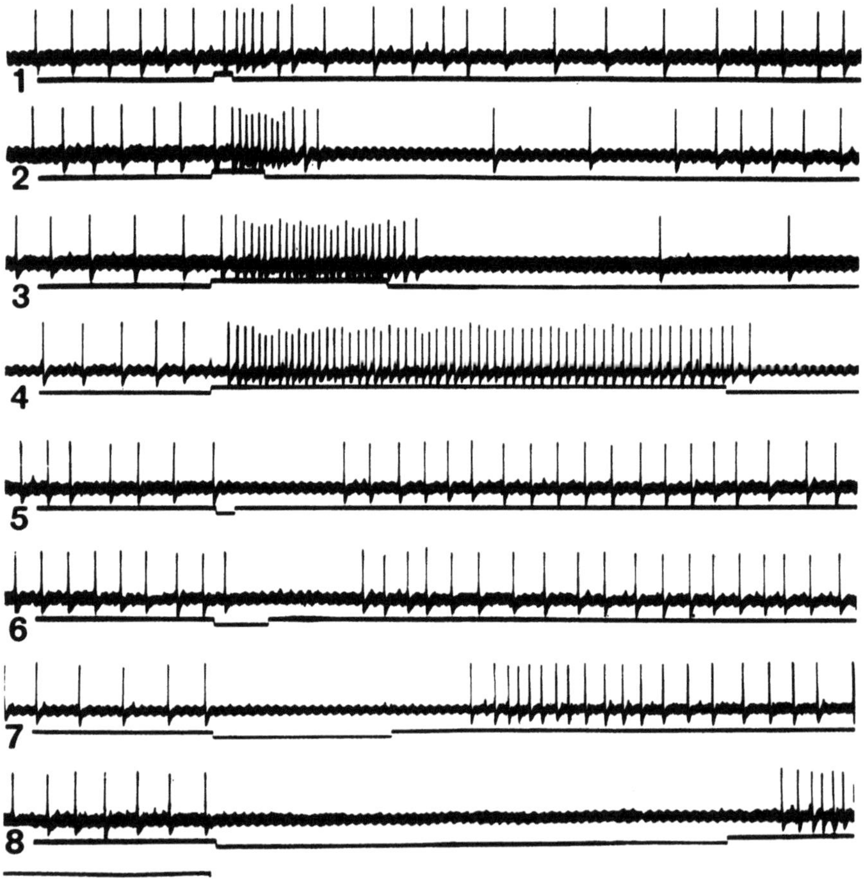

Fig. 30,1–8. Effect of stimulus duration and polarity on responses of the ampullae of Lorenzini of the skate *D. pastinaca*. **1–4** Responses to cathodic pulses **5–8** Responses to anodic pulses. Pulse durations are 40 ms, 120 ms, and 400 ms for **1**, **5**; **2**, **6**; and **3**, **7**, respectively. The pulse intensity is 6×10^{-9} A/mm^2, t = 17 °C. *Time bar* 500 ms. (Akoev et al. 1976a) (Reprinted with permission from Comp. Biochem. Physiol. v. 53A, Akoev et al. Physiological properties of electroreceptors of marine skates, Copyright 1976, Pergamon Press PLC)

It should be noted that the character of changes in the background impulse activity of the ampullae of Lorenzini depended upon the polarity of the electric pulses. When a constant-voltage cathodal pulse was applied near the ampullary canal pore, the impulse frequency in the fibers was increased during the pulse (ON-response) and decreased after its termination (Fig. 30). The opposite polarity resulted in a decrease in impulse rate during the pulse, and the after-effect was an increase in frequency (OFF response).

In a series of current- and voltage-clamp experiments on electrically isolated ampullae of Lorenzini in skates, a lumen-positive DC potential of 10–30 mV was recorded (Clusin and Bennett 1977a,b). Negative stimuli evoked a large lumen-negative receptor spike up to 100 mV high and lasting for 100 ms (Obara and Bennett 1972; Clusin and Bennett 1977a,b). Voltage-clamp experiments revealed four types of ionic currents in addition to the leakage type in the ampullae of Lorenzini of the skates: voltage-dependent Ca^{2+} currents of the apical and basal faces, Ca^{2+}-activated K^+ current of the apical face, and voltage-dependent K^+ current of the basal face. It is noteworthy that the Ca^{2+} current of the apical membrane acts only as the charge carrier dominating in the generation of the receptor spike. The basal Ca^{2+} current was much smaller and was involved in transmitter release.

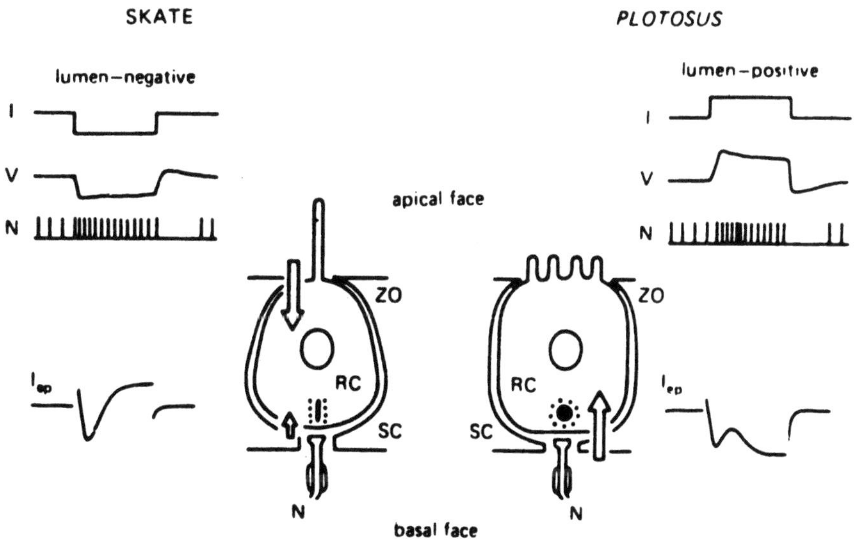

Fig. 31. Diagram of electroreceptor cells of nonteleost (skate) and teleost (*Plotosus*). *RC* receptor cell; *SC* supporting cell; *ZO* zonula occludens; *N* afferent nerve. Receptor cell membrane is divided by ZO into apical and basal faces. Main inward current components, Ca current, are shown by *thick arrows*. *Upper insets* in situ responses to facilitatory stimuli are shown schematically. *I* stimulus; *V* lumen potential positively shown upward; *N* single afferent response. Note different stimulus polarity for afferent facilitation. *Lower insets* epithelial current pattern. *Iep* active current, inward in respective membrane site shown *downward*. (Obara and Sugawara 1984)

This current is blocked by Ca^{2+}-free solution and by Co^{2+}, Sr^{2+} being a substitute for Ca^{2+}.

Such properties of the ampullae of Lorenzini have been suggested to be due to the excitability of the lumenal faces of the receptor cells (Obara and Bennett 1972). In the ampullae of Lorenzini, the regenerative response arises on the apical surface of the receptor cell (Figs. 31, 32) (Obara and Bennett 1972; Bennett and Obara 1986).

Negative stimuli applied to the canal pore depolarize the apical membrane, changing its permeability to Ca^{2+}, thereby giving rise to a negative response of sensory epithelium (Obara and Bennett 1972). Depolarization of the apical membrane induces current flows out through the basal membrane, depolarizing it and increasing the secretion of the neurotransmitter. The anodal stimulus, in contrast, hyperpolarizes the apical membrane and closes the Ca^{2+} channels, thus decreasing the inward current through this face, which leads to the hyperpolarization of the basal membrane and a reduced secretion of the transmitter.

According to another hypothesis (Brown and Govardovsky 1983a,b; Brown and Ilyinsky 1984), the unusual response of the ampullae of Lorenzini in elasmobranchs to the polarity of electric stimulation seems to be due to the presence of steady-state negative slope conductance of the receptor cell

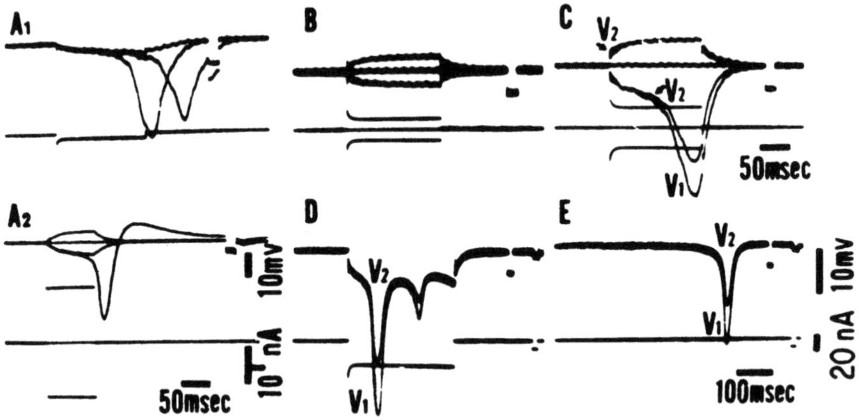

Fig. 32A–E. Receptor spikes evoked by cathodal stimulation in the ampullae of Lorenzini of the skate. *Upper traces* ampullary voltages; *upward deflections* lumenal positivity. *Lower traces* currents; those making the lumen positive shown as upward deflections. **A_1** and **A_2** Threshold stimulation of two preparations. **B** to **E** Recordings were made simultaneously near to (V_1) and far from (V_2) ampulla, with a current electrode placed farther down in the canal. **B** Subthreshold cathodal stimulation and equal anodal stimulation superimposed. **C** Supra-threshold stimulation only. The receptor spike is larger near the ampulla. **D** A large cathodal current evoked two successive receptor responses with progressively declining size. **E** Spontaneous appearance of a receptor spike larger in amplitude near the ampulla. Voltage and current calibrations in **E** apply to **A** and **B** to **D**. (Obara and Bennett 1972) (Reprinted from the Journal of General Physiolgy 1972, v. 60, pp. 539 by copyright permission of the Rockefeller University press)

apical membrane. On the contrary, the ampullary receptors of *Plotosus* respond to the polarity of electric stimuli in an opposite manner to the receptors of elasmobranchs (Fig. 31). The apical faces of hair cells appear to be electrically inexcitable, while the basal faces contain voltage-sensitive Ca^{2+} channels. In these receptors, stimulating current directly affects the basal membrane. If anodal stimulus is applied to the exterior of the receptor, the inner (presynaptic) face of the receptor is depolarized, thereby increasing the rate of transmitter release and the firing of afferent fibers.

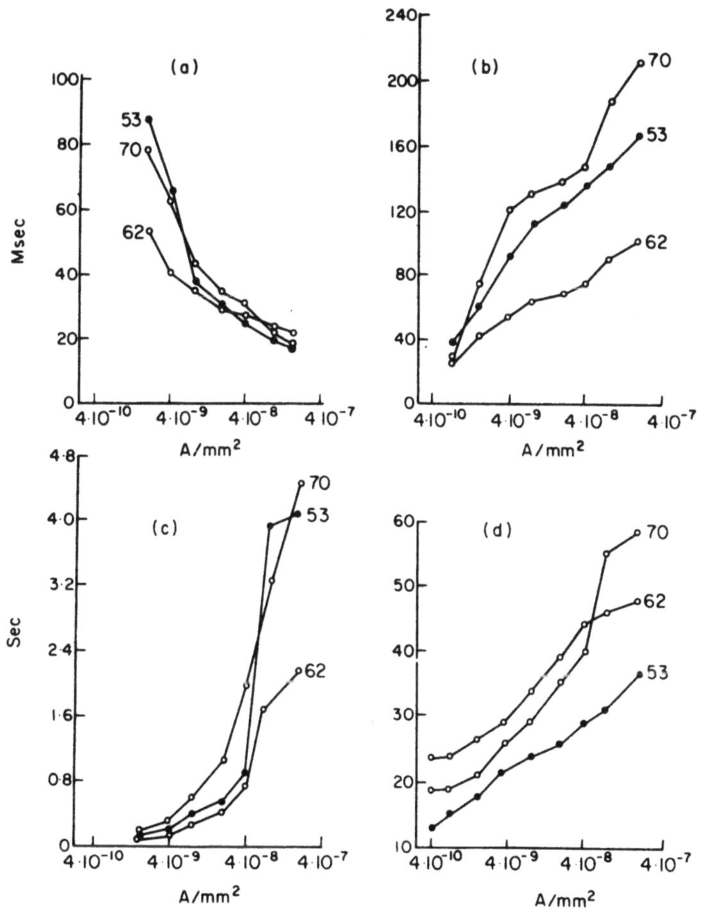

Fig. 33a–d. Effect of intensity of applied pulse on the responses of the ampullae of Lorenzini of the skate *D. pastinaca*. **a,b** Effect of intensity of pulse on latency of ON- and OFF-responses. **c** Relations between the intensity of applied pulses and duration of inhibition following the train of impulses. **d** Relations between the intensity of pulse and firing rate of OFF-response. The duration of the electric pulse is 300 ms, t = 16 °C. Numbers of fibers at the *right* of each curve. (Akoev et al. 1976a) (Reprinted with permission from Comp. Biochem. Physil. v. 53A Akoev et al. Physiological properties of electroreceptors of marine skates, Copyright 1976, Pergamon Press PLC)

The ampullae of Lorenzini in skates responded to electric pulse with phasic-tonic discharge. Negative stimuli increased their firing rate after the onset of a stimulus with gradual adaptation. Excitation was followed by a period of rebound inhibition. Latency in response to moderate stimuli varied from 20 to 40 ms for the ON-response, and 50–100 ms for the OFF-response. The latency of ON-responses varied inversely with stimuli strength whereas the latency of OFF-responses varied linearly (Fig. 33a,b). The minimal latency for the ON-burst to strong stimuli was about 15 ms.

A linear relationship was found to exist between current intensity and response rate. In some units an inhibition of the discharge was observed at a high intensity of current. The main effect of an increased intensity of the applied pulse was an increase in the OFF-response frequency and the duration of rebound inhibition following the termination of ON-responses (Fig. 33c,d). The time course of rebound inhibition following cathodal excitation increased with lengthening of the stimulus (Fig. 29a). An increase in the duration of imposed pulses produced an increase in the firing rate of ON- and OFF-responses (Fig. 29b). It is worth noting that the OFF-response latency was maximal for a stimulus of 80–100 ms. Both increase and decrease in stimulus duration produced a decrease in latency (Fig. 29c). The characteristics of rebound inhibition following the cathodal excitation were analyzed by means of two electrical pulses. The rebound inhibition disappeared when the interval between two pulses increased. The time course of this inhibition depends on the intensity and duration of the first pulse applied.

An important feature of many sensory receptors is their ability to adapt to long-lasting constant stimuli. This phenomenon is defined as a decrease with time in a response of receptors to a constant input. In the ampullary receptors, numerous investigations have demonstrated their adaptation to constant stimuli. The ampullae of Lorenzini in the skate can adapt to long-lasting constant current of opposite polarity and intensity up to the 10^{-7} A/mm^2. At the initiation of cathodal and the termination of anodal stimuli, a dynamic overshoot in frequency occurred, depending on the current strength, followed by a gradual restoration of firing rate to the initial level (Fig. 34a,b). On the contrary, the switching off of cathodal and the switching on of anodal current led to a transient inhibition of static discharge followed by a gradual reappearance of activity. The duration of transient phases of adaptation depended on the intensity of the stimuli applied. A change in the excitability of receptors in the course of adaptation to constant current was examined with short (50-ms) test stimuli. The latency of responses to test stimuli proved to decrease gradually; and the number of elicited impulses increased following the switching on of anodal and the switching off of cathodal current (Fig. 34c,d).

Thus, the important conclusion which can be drawn from these experiments is that the excitability of receptors adapting completely to the long-lasting constant current of any polarity does not differ from that prior to stimulation. Such a significant feature in the ampullae of Lorenzini enables

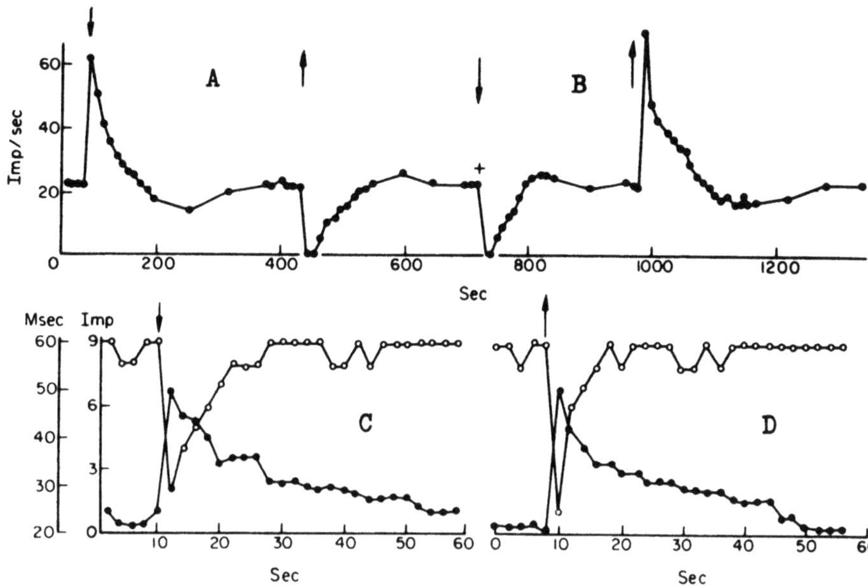

Fig. 34A–D. Adaptation of the ampullae of Lorenzini to long-lasting constant electrical current. **A** Time course of nerve impulse frequency following onset and termination of cathodal (**A**) and anodal (**B**) currents. The current intensity is $2 \times 10^{-7}\,A/mm^2$. **C,D** The changing of the latency (*filled symbol*) and the number of impulses (*open symbol*) per 50 ms test electric pulse after switching on the anodal (**C**) and switching off the cathodal (**D**) constant current of $2 \times 10^{-7}\,A/mm^2$. The test pulse intensity is $3 \times 10^{-7}\,A/mm^2$, t = 16 °C. *Arrows downward* and *upward* indicate the switching on and off of current, respectively. (Akoev et al. 1976a) (Reprinted with permission from Comp. Biochem. Physil. v. 53A Akoev et al. Physiological properties of electroreceptors of marine skates, Copyright 1976, Pergamon Press PLC)

them to maintain a high sensitivity to changing stimuli in the presence of constant potentials in surrounding water.

The process of adaptation of the ampullae of Lorenzini to long-lasting current may, in principle, take place in both the receptor cell and the afferent fiber. Both suggestions seem reasonable. However, the latter alternative can be entirely ruled out, since this assumption was based on the fact that receptors adapted to anodal current when there were no impulses in the afferent fibers. Several lines of evidence indicate that the onset of cathodal current depolarized the receptor cell and evoked a burst of activity in afferent fiber. During the adaptation of the receptor, the threshold and firing rates were restored. This depolarization may be assumed to be compensated by the potential (e.g., active transport) generated in the sensory cell itself. If this is the case, the opposite changes in the impulse activity should appear following the termination of the current actually observed in experiments. Anodal current has the opposite effect. Hence, for long-lasting stimuli, response patterns are identical at the switching on of

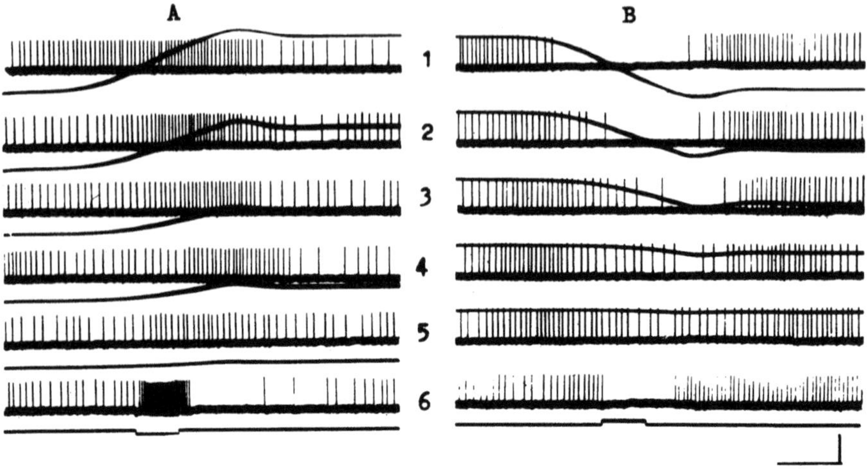

Fig. 35A,B. Single unit responses of dorsal ampullae of the skate, *Raja clavata*, to magnetic stimuli of various intensities and opposite directions. **A,B** Northern and southern directions of the field, respectively. Intensity of magnetic stimulation in 10^{-4} T/s: *1* 67; *2* 53; *3* 17; *4* 8; *5* 3.3. *Solid trace* represents the changing magnetic induction. *6* Responses of the same receptor to cathodal (**A**) and anodal (**B**) current intensity of 3 10^{-9} A/mm^2. *Vertical bar* 3 × 10^{-3} T/s; *time bar* 500 ms. (Akoev et al. 1976b)

cathodal and at the switching off of anodal currents. For the same reason, the effects are identical at the switching off of cathodal and the switching on of anodal currents of equal strength. A growing body of evidence indicates that in some receptors, and perhaps in the ampullae of Lorenzini, the processes of active transport can play an important role in adaptation to maintained stimuli (Sokolove and Cooke 1971).

The high sensitivity of the skate ampullae of Lorenzini to electric stimuli suggests a possibility of perceiving the Earth's magnetic field (Kalmijn 1974, 1982). In electrophysiological experiments, responses to the magnetic field were shown in the ampullae of Lorenzini (Akoev et al. 1976b; Brown and Ilyinsky 1978) and in central neurons (Andrianov et al. 1974). The receptors proved to respond to a changing magnetic field only (Fig. 35), and the character of the responses depended on both the direction of the magnetic field and the position of the receptor. The sensitivity of receptors to the magnetic field was found to depend on the length of the ampullary canal. More recently, responses of the ampullae of Lorenzini to the Earth's magnetic field were demonstrated (Brown et al. 1979). Hence it follows that sharks and skates may use the Earth's magnetic field for orientation and navigation. A detailed study of electroreceptor response to magnetic stimuli was made by Brown and Ilyinsky (1984).

Important characteristics of ampullary electroreceptors were found using the current- and voltage-clamp technique (Obara and Bennett 1972; Clusin and Bennett 1977a,b; 1979a,b; Obara 1976; Bennett and Obara 1986). The

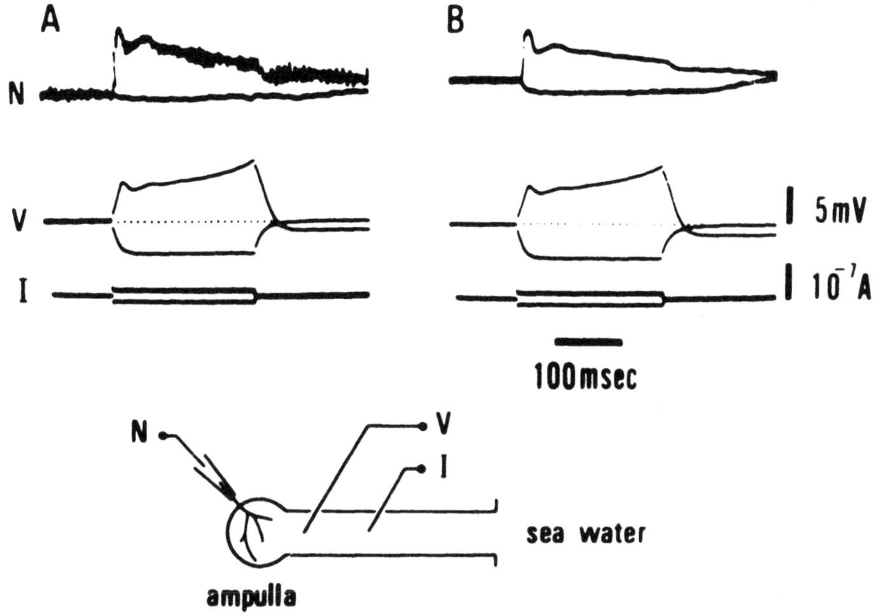

Fig. 36A,B. Examples of ampullary electroreceptor responses in situ of the marine catfish *Plotosus*. *Lower diagram* experimental set-up. *N* afferent nerve activity recorded by a suction electrode; *V* microelectrode recording of responses in the lumen of ampullae (sensory epithelium); *I* constant current stimulation. *Upper records* superimposed responses (negativity for I and depolarization for V and N shown *upward*). **A** Note small damped oscillation (*V*) and depolarizing PSP with marked increase of afferent impulses (*N*) by lument-positive pulse (*I*). Stimuli, supramaximal for the afferent output, evoked late grated regenerative response in *V*. **B** TTX (10^{-6} g/ml) abolished nerve impulses revealing underlying PSP, but left the ampullary response unaltered. (Sugawara and Obara 1984a) Reprinted by permission from Brain Research v. 302, pp. 172, 1984)

available literature contains detailed data on the primary processes in the ampullae of Lorenzini of the marine catfish, *Plotosus* (Obara and Oomura 1973; Obara 1976; Obara and Sugawara 1979; Sugawara and Obara 1984a,b, 1989; Bennett and Obara 1986; Sugawara 1989a). Investigation on *Plotosus* electroreceptors in situ showed that without stimuli the initial DC level in the ampullae was 0–3.1 mV. Lumen-negative stimuli induced a hyperpolarizing response in the ampullae with the cessation of impulse discharges in the afferent fibers (Fig. 36A). Lumen-positive stimuli induced a depolarizing ampullary response with damped oscillation (Fig. 36A). An increase in the intensity of stimuli evoked a regenerative response due to the entry of Ca^{2+} into the basal membrane. The amplitude of the regenerative response was determined by the external Ca^{2+} concentration. It is worth noting that ampullary response was unaffected by tetrodoxin (Fig. 36B). In electrically isolated ampullae there was a DC negative potential of 20 mV inside the ampullae, in contrast to the DC potential in situ, which was near zero.

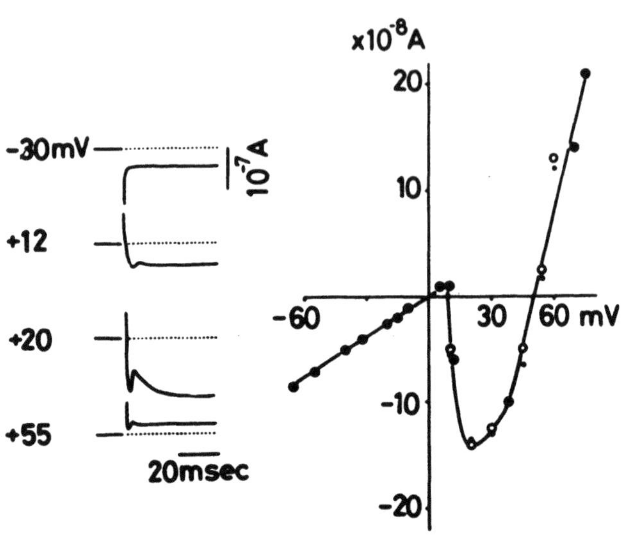

Fig. 37. Ionic currents in *Plotosus* ampulla. A family of current records on different voltage steps marked is shown *leftward*. The voltage levels are measured from holding potential set at the resting DC potential of the ampulla. Both early and late inward currents are near their maxima at +20 mV. Their peak values are plotted in the V-I relations shown *rightward*. *Open circles* early currents. *Small dots* late currents. When both reach the same value *small dots are placed in open circles*. Note their identical negative impendance characteristics. (Obara and Sugawara 1979)

Positive stimuli evoked a regenerative response which considerably outlasted the duration of the stimulus. As in the case of freshwater fish electroreceptors, the apical membrane of the *Plotosus* receptors seems to be electrically inexcitable, while the basal membrane contains voltage-sensitive channels mediating the release of transmitter. The receptors possess resting tonic discharges which are modulated by ±10% by stimuli of a few microvolts.

Voltage clamp experiments showed two distinct components in both inward and outward current (Obara and Sugawara 1979; Sugawara and Obara 1989; Sugawara 1989a) (Fig. 37). The inward currents were insensitive to TTX and comprised an early transient component with a peak at 50 ms. The inward currents reverse their sign at 34 mV. Both of the inward components are carried by Ca^{2+} in the basal membrane of the receptor cells, since in Cd^{2+} solution these currents are completely abolished. It is worth noting that the apical membrane seemed to be inactive when the external Ca^{2+} was increased from 3 to 10 mM, the inward current increased, and the reversal potential was shifted from +35 to +47 mV. The permeability of the channels to divalent cations was of the order of Sr^{2+}>Ca^{2+}>Ba^{2+}. Divalent cations (Cd^{2+}, Co^{2+}, Mn^{2+}, and Mg^{2+}) reduced the inward current. Verapamil in a concentration of 20 mM suppressed the inward current by 28% (Sugawara and Obara 1989). The above data have revealed two components of the Ca^{2+} current. In normal solution, Ca^{2+} current turned out to decay in two

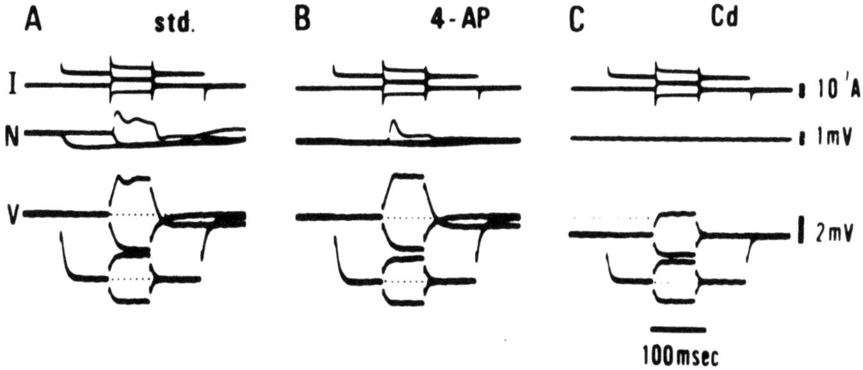

Fig. 38A–C. Suppression of damped oscillation and V-dependent nonlinear response by K^+- and Ca^{2+}-blockers, respectively, in *Plotosus* ampullary electroreceptors. *Each record* four superimposed responses to short test pulses with or without longer conditioning hyperpolarizing pulse. *I* constant current stimulation. *N* afferent nerve activity recorded by a suction electrode. *V* microelectrode recording of responses in the lumen of ampullae (sensory epithelium). *Upper records* superimposed responses to successive current pulses (negativity for I and depolarization for V and N shown *upward*). **A** Note damped oscillation (*V*) at control DC level and V-dependent nonlinearity of ampullae. **B** K^+-blocker, 2 mM 4-aminopyridine (4-AP) abolished oscillation but asymmetrical nonlinear responses remained at the control DC level. **C** Addition of Ca^{2+}-blocker, 0.1 mM cadmium (Cd^{2+}) further abolished nonlinearity and shifted the DC level to a negative one. Transmission to the afferent nerve was completely abolished (*N*). (Sugawara and Obara 1984a) Reprinted with permission from Brain Research v. 302, pp. 172, 1984)

exponential phases (Sugawara 1989a). The fast Ca^{2+} current was sensitive to external Na^+ and abolished in low Na^+ solution. This current seems to activate K^+ current and the phasic release of transmitter. On the other hand, the slow component of Ca^{2+} current is less sensitive to external Na^+ and inactivated to a lesser degree. This component seems to contribute to the modulation, under the action of external stimuli, of the tonic receptor operation and the continual transmitter release due to the depolarization of sensory epithelium. Thus, these results suggest two populations of the Ca^{2+} channels in the basal membrane of the electrosensory cell of *Plotosus*. The outward current seems to be carried out by K^+ ions (Sugawara and Obara 1984a). The blockers of K^+ channels such as 4-AP or TEA blocked the outward current (Fig. 38B). On the other hand, this current was also abolished by Co^{2+} and Cd^{2+}, suggesting a transient Ca^{2+}-activated K^+-current (Fig. 38C). The oscillation of receptors in *Plotosus* (Sugawara and Obara 1984a,b) and *Raja* (Clusin and Bennett 1979a,b) was due to the modulation of the inward current of the apical membrane through the interaction with Ca^{2+} and Ca^{2+}-activated K^+ conductance of the basal face. In the ampullae of Lorenzini of the Skate *Raja*, a lumen-negative stimulus produced a small increase in oscillation (Clusin et al. 1975. Clusin and Bennett 1979a). These oscillations were blocked by Co^{2+} and Cd^{2+}. In the basal membrane of *Plotosus* sensory cells contain Ca^{2+}-activated K^+ channels which seem to contribute to the initial phase of receptor adaptation. The perfusion of the basal membrane with caesium (10–20 mM) solution has been shown to result in a reduction of adaptive characteristics of receptors (Brown et al. 1985). Moreover, the K^+ conductance may also be essential to a stable operation of the extremely sensitive receptors.

2.3.2 Temperature Sensitivity of Electroreceptors

Ampullary electroreceptors are known to be highly sensitive to temperature stimuli (Sand 1938; Hensel 1955, 1974; Pierau and Meissner 1966; Roth and Szabo 1970; Bromm et al. 1975; Akoev and Volpe 1978; Akoev and Zhadan 1978; Akoev et al. 1980d; Wissing et al. 1988; Akoev 1990). For a long time these receptors were considered as typical cold thermoreceptors since their responses were similar to those of cold receptors in mammals. Their sensitivity is indeed comparable to that of mammalian cold receptors. Thus, the sensitivity of the ampullae of Lorenzini is 40 imp/s/°C for *Raja* (Sand 1938), and −90 imp/s/°C, for sharks (Hensel 1955), and that of the cat's cold

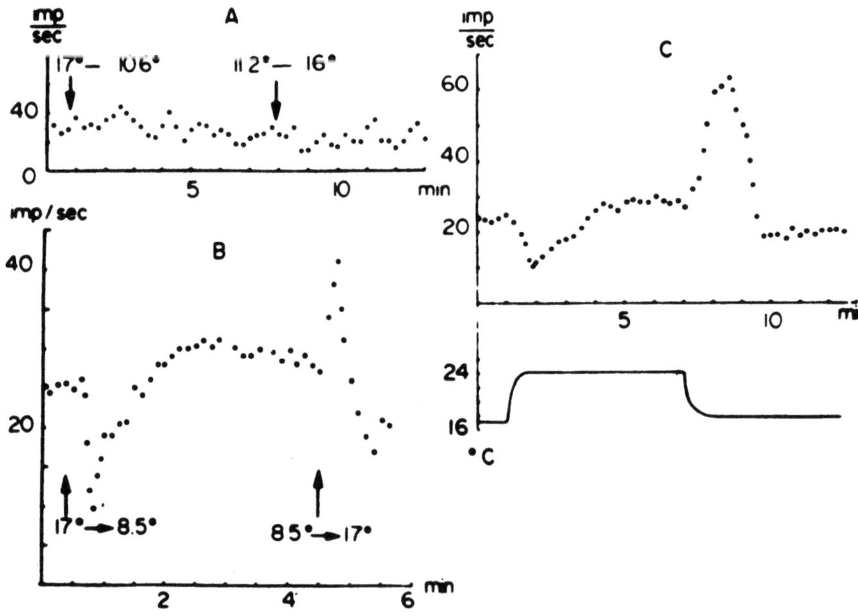

Fig. 39A–C. Effect of temperature change on the impulse frequency of the ampullae of Lorenzini of the skate *R. clavata*. Effect of cooling and warming of: **A** canal opening; **B** a part of the ampullary canal; **C** the receptor cells. *R. clavata* dorsal receptor with 7-cm-long ampullary canal. (Akoev et al. 1980d) (Reprinted with permission from Comp. Biochem. Physiol. v. 65A, Akoev et al. Analysis of effect of chemical and thermal stimuli on the ampullae of Lorenzini of the skates. Copyright 1980, Pergammon Press PLC)

receptors is −30 to −55 imp/s/°C (Hensel 1955). The sensitivity of the ampullae of Lorenzini to temperature stimuli was as low as 0.05 °C.

Electroreceptor responses to temperature stimulation were characterized by a negative temperature coefficient, i.e., cooling caused an increase in the firing rate (dynamic phase) followed by adaptation to the level lower than the one preceding the temperature change (Fig. 39C). On the contrary, during warming, the reverse effect was observed. The static firing rate has a positive temperature coefficient and it increased as the temperature was raised from 4 to 25 °C and decreased again at a temperature above 25 °C (Hensel and Nier 1971; Bromm et al. 1975; Nier et al. 1976; Akoev et al. 1980d). However, the mechanism of the high temperature sensitivity of electroreceptors is still obscure. As the ampullae of Lorenzini are secondary receptors, possible sites of thermal stimuli action may be as follows: (1) the apical membrane of a sensory cell; (2) the basal membrane involved in transmitter replease; (3) postsynaptic membranes of afferent nerve fibers; (4) changes in electrical properties of the canal wall or in the conductivity of the canal jelly.

Some authors assumed that changes in temperature mainly affect the presynaptic mechanism involved in transmitter release (Machin 1962). However, the mechanisms involved in transmitter release and the postsynaptic potential amplitude are known to have a positive temperature coefficient. It is therefore possible that the static temperature component of the ampullae of Lorenzini is due to the effect of temperature on the prejunctional membrane involved in the transmitter release, while the dynamic component results from the effect of temperature on the apical face of the receptor cell. This suggestion is supported by the evidence that the receptors with low electric sensitivity failed to exhibit the dynamic component although static firing rate changes did occur.

Different sites of action of thermal stimuli are discussed in the literature, among them the postsynaptic membrane (Murray 1962b). Experiments on the ampullae of Lorenzini revealed the absence of high thermal sensitivity of the postsynaptic membrane because the ampullae showed a great loss of sensitivity, and inverted thermal responses were observed in a solution containing an elevated Mg^{2+} concentration when the firing of the afferent fibers was restored by L-ASP or L-GLU (Fig. 40).

Murray (1974) believed that the ampullary canals or the openings were not involved in the processes of temperature sensing by the receptor. The dorsal ampullae of Lorenzini in *Raja*, which have a long ampullary canal, are an appropriate preparation to evaluate to what extent the ampullary canal contributes to temperature transduction (Akoev et al. 1980d). Experiments in vivo showed that a temperature change at the canal opening produced no effect on the firing rate (Fig. 39A). Cooling part of the canal (2–4 cm) resulted in a transient inhibition of the firing rate followed by adaptation to the initial level, whereas warming elicited a dynamic overshoot in frequency (Fig. 39B). This phenomenon may be explained by a change in

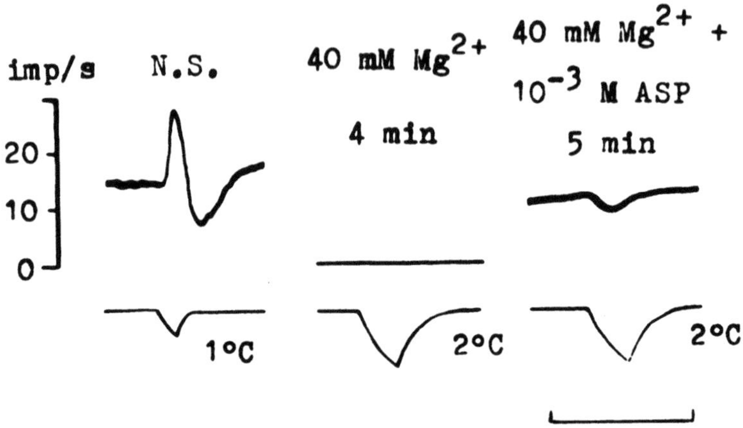

Fig. 40. Inverted temperature responses of the ampullae of Lorenzini in high-Mg^{2+} solution with aspartic amino acid. *Time bar* 2 min. (Akoev 1990)

the electric parameters of the canal wall or in the conductivity of the jelly. These responses were the inverse of those obtained by thermal stimulation of the receptor cells (Fig. 39C). Thus they indicate that thermal changes in the ampullary canal counteract rather than contribute to the normal thermal responses.

It is worth noting that most investigators are inclined to believe that the sensory epithelium is thermosensitive (Roth and Szabo 1970; Bromm et al. 1975; Akoev et al. 1980d; Akoev 1990). Several lines of evidence indicate that during thermal stimulation the membrane potential of sensory and supporting cells is changed, which is perceived by electroreceptors as an adequate stimulus. During a simultaneous recording of the ampullary

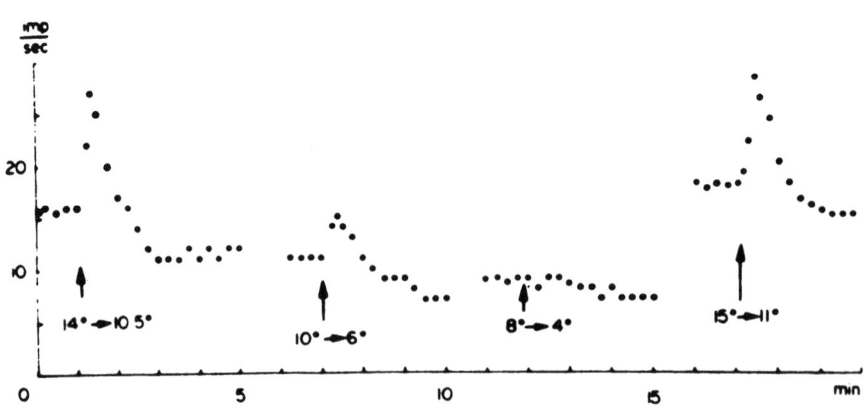

Fig. 41. Effect of stationary temperature level on the responses to cooling in the ampullae of Lorenzini of the skate. (Akoev et al. 1980d) (Reprinted with permission from Comp. Biochem. Physiol. v. 65A, Akoev et al. Analysis of effect of chemical and thermal stimuli on the ampullae of Lorenzini of the skates. Copyright 1980, Pergamon Press PLC)

transepithelial potential and the impulse activity of the nerve fibers, it was found that warming of the sensory epithelium evoked a negative potential at the basal face and a suppression of the firing rate (Brown and Govazdovsky 1981, 1982).

The dynamic responses consist of a transient frequency overshoot on cooling and a transient inhibition or reduction of the impulse activity on warming. The amplitude of the dynamic frequency change depends on the temperature at which temperature change was introduced. The lowering of stationary temperature resulted in a decrease in dynamic responses. Below 8 °C, they were sometimes completely blocked (Fig. 41).

Further information with respect to the possible sites of sensory tranduction in the ampullae of Lorenzini was obtained when comparing the response characteristics during thermal and electrical stimulation. The electrically induced firing changes were found to depend on the actual temperature. The electric current threshold was temperature-dependent, and over the range of 7–25 °C it changed by a factor of 10. The receptors appeared to be most sensitive to electric current within the temperature of 12–18 °C. Also, the number of impulses produced by an electric pulse increased with rise in temperature (Fig. 42), while the latency of responses was found to increase as the temperature decreased. Additionally, it is

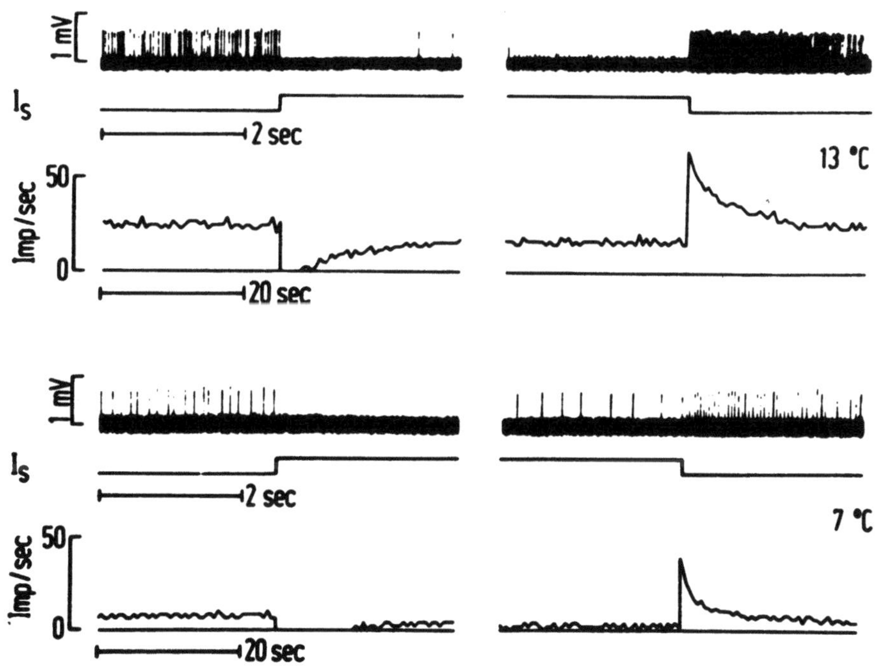

Fig. 42. Responses of the ampullae of Lorenzini of shark to electric current at different temperatures. *Upper line in each series* neural response patterns; *middle line* electrical current, hyperpolarizing current *upward*; *lower line* computer evaluation of impulse rate (averaged for every half s). (Bromm et al. 1975)

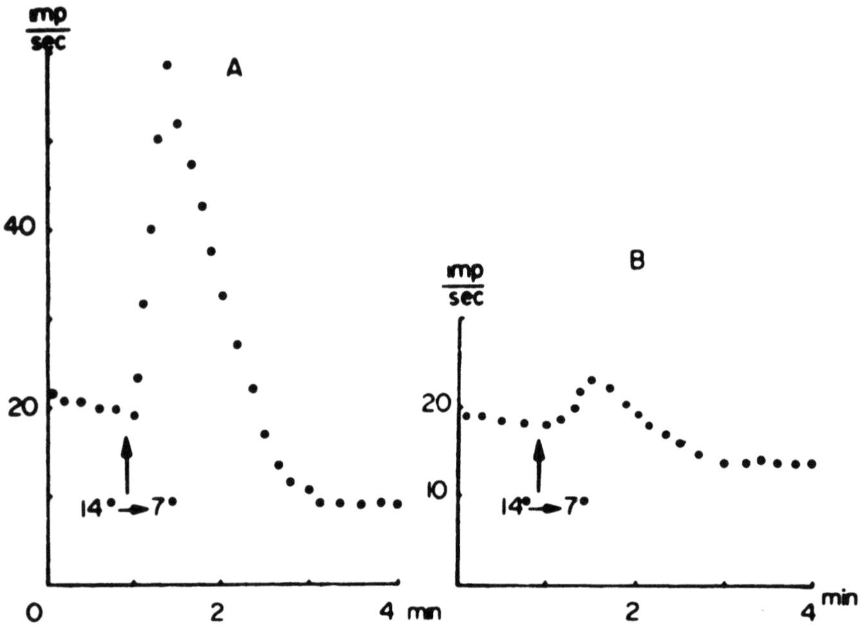

Fig. 43A,B. Time course of nerve impulse frequency responses to cooling of two ampullae of Lorenzini of skate with different electric thresholds. **A** 1.5×10^{-10} A/mm^2. **B** 1.0×10^{-8} A/mm^2. (Akoev et al. 1980d) (Reprinted with permission from Com. Biochem. Physiol. v. 65A, Akoev et al. Analysis of effect of chemical and thermal stimuli on the ampullae of Lorenzini of the skates. Copyright 1980, Pergamon Press PLC)

interesting that the thermal sensitivity of the receptors depended on electric threshold and vice versa (Fig. 43). Therefore, in both cases a common mechanism is likely to underlie the receptor sensitivity, e.g., a change in the membrane potential of the receptor cell. Murray (1959) is of the opinion that a thermal stimulation is equivalent to an electrical one.

Adaptation of the ampullae of Lorenzini to long-lasting constant electric current at various temperatures showed that although the dynamic overshoot in firing rate decreased with lowering temperature, the time course of adaptation did not change significantly. It is worth noting that at a low temperature the responses to a break-off of cathodal and switch-on of anodal currents were poorly pronounced due to a low level of resting discharges.

It is increasingly evident that Ca^{2+} is essential to various processes in sensory cells. Perfusion experiments on the ampullae of Lorenzini showed that changes in the external Ca^{2+} concentration significantly affect the temperature sensitivity of receptors (Akoev and Andrianov 1981; Akoev 1990). In a high Ca^{2+} solution the thermal sensitivity of receptors decreased. When Ca^{2+} was completely removed from the external solution, the thermal and electrical sensitivity initially decreased and then increased (Fig. 44). If

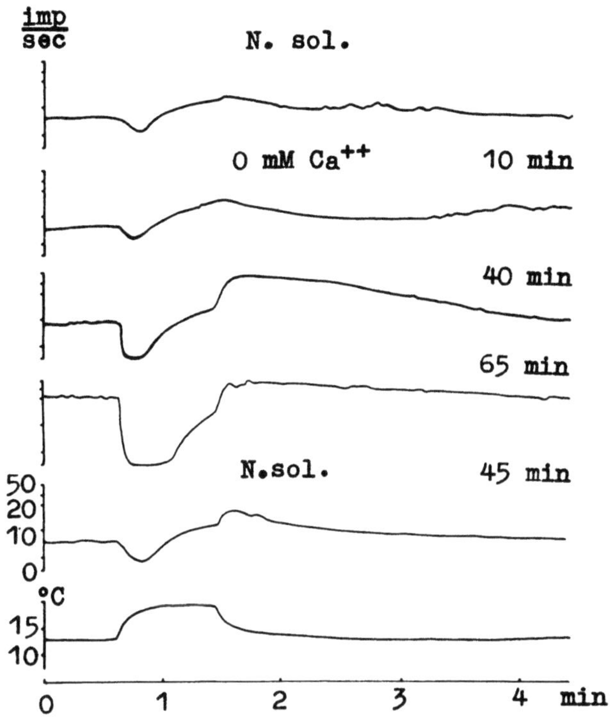

Fig. 44. Effect of Ca^{2+}-free solution on responses of the ampullae of Lorenzini of the skate to thermal stimuli. Response in normal solution; responses in Ca^{2+}-free solution (10, 40, and 45 min, respectively); recovery in normal solution (Akoev and Andrianov 1981)

perfusion was prolonged, the responses to thermal stimuli were abolished and reappeared in normal solution. These results might be due to different sites of Ca^{2+} action which are known to affect not only transmitter release but also the control of ionic membrane channels. For example, increasing Ca^{2+} concentration produces (1) an increase in Ca^{2+} equilibrium potential, E_{Ca}, which promotes Ca^{2+} influx and a consequent rise in transmitter release, and (2) an increased electrostatic screening of the negative surface charges near Ca^{2+} channels, which causes hyperpolarization across the channels and depresses the Ca^{2+} influx. The opposite effects take place when Ca^{2+} concentration is reduced. In a Ca^{2+}-rich solution, the net Ca^{2+} influx is possibly decreased due to hyperpolarization across the Ca^{2+} channels and therefore the thermal sensitivity of receptors is depressed. The phase of a decrease in the thermal sensitivity of the ampullae of Lorenzini in a Ca^{2+}-free solution appeared to be due to a decrease in E_{Ca}, whereas the phase of an increase in excitability may be explained by an effective depolarization across the voltage-sensitive Ca^{2+} channels which became more sensitive to small changes in the membrane potential.

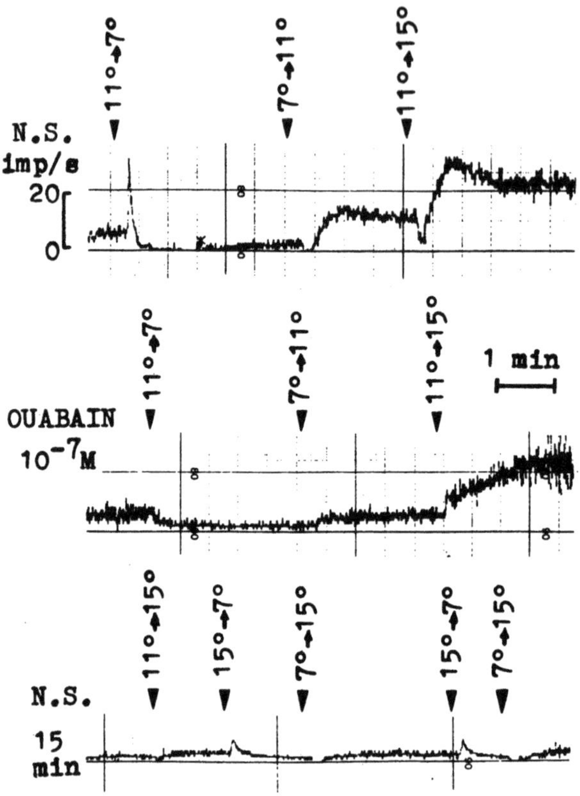

Fig. 45. The effect of ouabain on thermal responses of the ampulla of Lorenzini of shark. *N.S.* normal solution (Akoev 1990)

Neuronal membrane potentials are known to depend upon ionic diffusion potentials and upon an electrogenic pump. There is some reason to believe that the electrogenic pump can also contribute to the resting potential of electroreceptor cells. Investigations on *Aplysia* neurons have demonstrated that cooling results in their depolarization by up to 35 mV, and this effect can be blocked by ouabain, a pump-inhibiting agent (Carpenter 1970). These neurons are characterized by a very high specific membrane resistance and a high temperature dependence of the active transport process. Hence, a small temperature change is capable of evoking a sizeable potential in neurons. Warming of *Aplysia* neurons has been shown to lead to hyperpolarization, and this is converted into depolarization if the electrogenic pump is blocked by ouabain (Junge and Ortiz 1978). In a voltage-clamped neuron, warming from 7 to 22 °C evoked an electrogenic pump current of 16 nA.

Several lines of evidence indicate that an electrogenic pump is likely to be the mechanism responsible for the high temperature sensitivity of specific cold receptors (Pierau et al. 1974; Spray 1974). As electroreceptor cells in skates have a high membrane-specific resistance and a high sensitivity to potential changes, an electrogenic pump might also contribute to the temperature sensitivity of the ampullae of Lorenzini (Akoev et al. 1980d). For example, temperature lowering may cause a decrease in pump activity, resulting in depolarization of the sensory cells and in a burst of impulse activity. Indeed, experiments on electroreceptors of sharks demonstrated that ouabain reduced or completely blocked the dynamic temperature responses (Fig. 45). The dynamic temperature sensitivity of the ampullae of Lorenzini, which decreases with a lowering temperature, thus might be due to a reduction in the contribution of electrogenic sodium pumping to membrane potential. In crayfish stretch receptor neurons, active sodium transport falls by 50% at 9 °C (Sokolove and Cooke 1971).

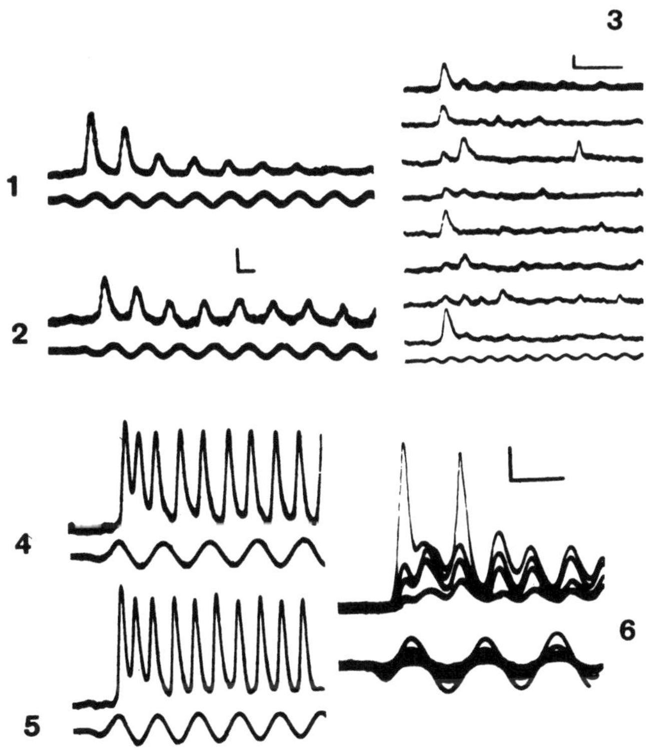

Fig. 46,1–6. Excitatory postsynaptic potentials of the sacculus of the goldfish. **1,2** Adaptive rundown of EPSPs (*upper traces*). Sound frequency 300 Hz (*lower traces*). *Vertical bar* 10 mV; *horizontal bar* 2 ms. (Furukawa and Matsuura 1978). **3** Fluctuation in the EPSPs to low intensity sound stimulation. *Upper trace* averaged record of 21 trails; *lower traces* individual records. Sound frequency (*lower traces*) 640 Hz; *vertical bar* 1 mV; *horizontal bar* 4 ms. (Furukawa et al. 1978). **4–6** Responses of auditory nerve fibers to different sound frequences. Generator potentials are evoked by different sound intensities. *Vertical bar* 10 mV; *horizontal bar* 2 ms. (Furukawa and Ishii 1967)

2.4 Electrical Responses of the Postsynaptic Membrane

The excitatory postsynaptic potentials (EPSPs) in the afferent fibers of the acousticolateral system were first studied experimentally on the goldfish sacculus (Furukawa and Ishii 1967). In nerve fibers, both action and generator potentials appeared to exist [Fig. 46(3–6)]. Later, the EPSPs were recorded itracellularly from the unmyelinated endings of the lateral-line afferents in the turbot *Lota lota* (Flock et al. 1973a,b). In the absence of mechanical stimulation, spontaneous EPSPs up to 5 mV were recorded from the afferent

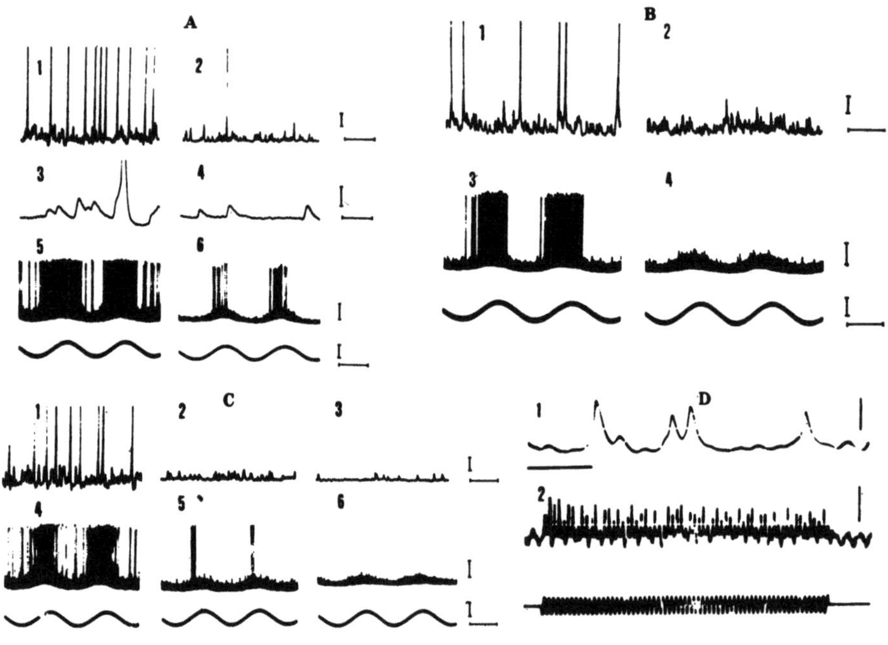

Fig. 47A–D. Responses of the hair cells of the acousticolateral system. **A** Intracellular recordings at rest and during stimulation from afferent nerve fibers of the semicircular canal in the frog. *1–4* resting activity in two units at slow and fast sweep speed. *Vertical bar* 10 mV; *horizontal bar* 100 ms. *5, 6* responses of the same units to two cycles of sinusoidal stimulation, 0.1 Hz, 10–110 deg/s. The turntable angular velocity is also indicated. *Vertical bar* 20 mV (*upper*), 100 deg/s (*lower*); *horizontal bar* 5 s. (Rossi et al. 1977). **B** Action of TTX (10^{-7} g/ml) (*2,4*) on resting (*1,2*) and evoked discharges (*3,4*) from afferent nerve fibers of the semicircular canal in the frog. Calibration is the same as in **A** (Rossi et al. 1977). **C** Effect of different Ca^{2+} and Mg^{2+} concentrations on resting (*1–3*) and evoked (*4–6*) activity of semicircular canal in the frog. *1,4* responses in 1.8 mM Ca^{2+} and O mM Mg^{2+}; *2,5* in 1.8 mM Ca^{2+} and 10 mM Mg^{2+}; *3,6*: in 0.9 mM Ca^{2+} and 10 mM Mg^{2+}. Calibration is the same as in **A**. (Rossi et al. 1977). (A,B,C Reprinted with permission from Brain Research v. 135, pp. 70, 71, 1977). **D** Excitatory postsynaptic potentials occur spontaneously in the afferent terminal of lateral line receptors of *Lota lota* (*1*). They synchronize in phase when the organ is mechanically stimulated (2) with a 70–Hz tone (*lower trace*). *Horizontal bar* 10 ms; *vertical bars*, 2 mV. (Flock et al. 1973a)

fibers (Fig. 47A,D). The synaptic delay for evoked EPSPs was 0.9–1 ms at 14 °C. Spontaneous EPSPs were also recorded from goldfish auditory fibers (Kyogoku et al. 1986) and hair cells of the lateral-line organ of *Xenopus laevis* (Sewell 1990). The latency of sound-evoked EPSPs was 0.4–0.8 ms.

A detailed analysis of EPSP characteristics was carried out in a series of studies on the goldfish sacculus (Ishii et al. 1971; Furukawa et al. 1972a,b, 1978, 1982; Furukawa and Matsuura 1978; Furukawa 1986). Statistical analysis of the transmitter release at the hair cell afferent fiber synapse in the sacculus of the goldfish has demonstrated that binomial rather than Poisson statistics fit better for a description of EPSP amplitude distribution. In response to each sound wave, the EPSPs show a marked adaptive rundown (Furukawa et al. 1978; Furukawa and Matsuura 1978) [Fig. 46(1,2)]. A similar phenomenon has also been observed in the lateral-line hair cells of *Xenopus* (Flock and Russell 1976).

EPSPs successively evoked in response to each sound wave, i.e., the size of the mean quantal content (m) were reduced by a fixed coefficient with respect to the preceding ones, irrespective of sound intensity, indicating that the rundown is attributable to a depletion of the available transmitter store. Thus, the adaptive rundown of EPSPs was found to be associated with a reduction in the size of the readily available store (n) but not with the probability of release (p). The rate of the rundown of successive EPSPs remained almost unchanged when the intensity of sound was changed. In those cases where the EPSPs completely adapted to a continuous sound, an increase in the stimulus intensity immediately caused EPSPs. Furukawa et al. (1978) regard this phenomenon as one of the forms of sensory adaptation based on transmitter depletion.

The hair cells were supposed to have many transmitter release sites with different sensitivities for release (Fig. 48). If the hair cells were stimulated

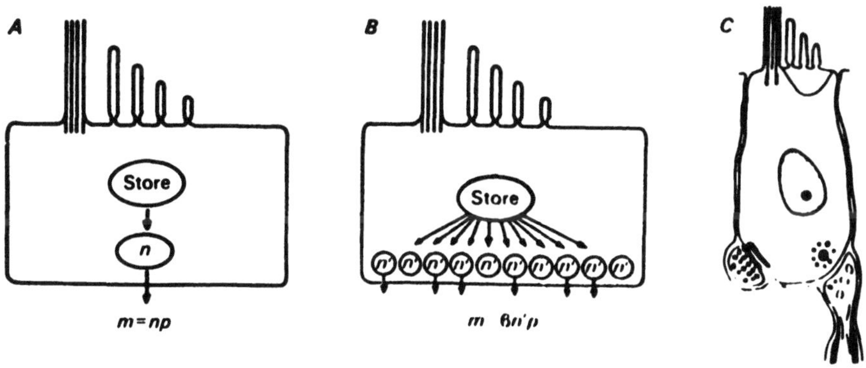

Fig. 48A–C. Hair cell models and their correlation with the morphology of hair cell afferent fiber synapse. **A** Single release site model **B** Multiple release site model **C** Schematic drawing of hair cell. The afferent synapse is characterized by the presence of a dense body surrounded by synaptic vesicles on the hair cell side. Several such sites exist within a single hair cell. (Furukawa and Matsuura 1978)

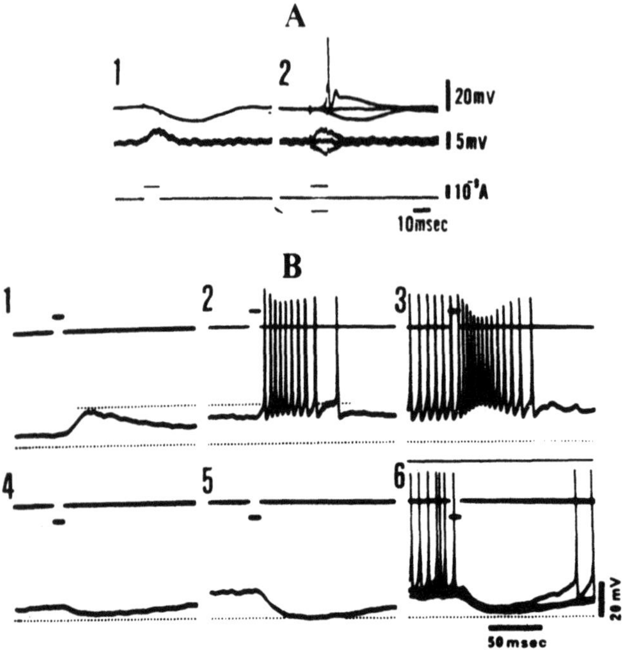

Fig. 49A,B. Electrical responses of afferent nerve fibers of electroreceptors. **A** Generator potential and spike in the afferent fibers of the ampulla of Lorenzini of the skate, *R. clavata*. *1* an anodal stimulus evokes a hyperpolarization of the nerve; *2* superposition of equal amplitude anodal and cathodal stimuli evoking hyperpolarization and depolarization, respectively. *Upper trace* intraaxonal recording; *middle trace* ampullary potential; *bottom trace* stimulating current. (Obara and Bennett 1972). (Reprinted from the Journal of General Physiology 1972, v. 60, pp. 545 by copyright permission of the Rockefeller University Press) **B** Intracellular recordings frome afferent terminal of electroreceptor of *Plotosus*. Repeated short current pulses induced the generator potential (*1*) and afferent discharges (*2*,*3*). *4*,*5* hyperpolarizing generator potential; *6* suppression of afferent discharges during hyperpolarizing generator potential. The *upper and lower dotted lines* show the firing level and the most negative membrane potential, respectively. (Reprinted from Electrobiology of nerve, synapse and muscle. Eds. Reuben JP, Purpura DP, Bennett MVL, Lendall ER. Obara S. Mechanism of electroreception in ampullae of Lorenzini of the marine catfish Plotosus. 1976, p. 134. With permission from Raven Press)

by a weak sound, the transmitter release would occur from a few particular sites with a probability of p being independent of the stimulus intensity. An increase in sound intensity would bring more and more release sites into activity.

The studies of synaptic transmission in the hair cells of the frog semicircular canal at rest revealed the presence of TTX-insensitive postsynaptic potentials of 0.5–10 mV, 3–6 ms duration, with a frequency of 100–300/s (Rossi et al. 1977, 1989) (Fig. 47A,B). Summation of these potentials led to the generation of action potentials.

PSPs were also recorded in the afferent fibers of ampullary electroreceptors in response to electric stimulation (Loewenstein 1960; Obara and Bennett 1972; Steinbach 1974; Obara 1976) (Fig. 49). In the ampullae of Lorenzini of *Raja*, a cathodal current pulse evoked slow depolarization in the axon (Obara and Bennett 1972). This depolarization potential clearly represented EPSPs. Their generation was accompanied by an increase in the conductance of the postsynaptic membrane. On increasing stimulus intensity, the EPSP gradually increased until it initiated spikes. The latency of EPSPs in response to stimuli was 10–20 ms. On the contrary, the positive stimuli evoked hyperpolarizing potentials in afferent fibers. Apparently, they cannot be regarded as true IPSPs, since they are not associated with the release of an inhibitory transmitter. They are, however, a consequence of repolarization of the postsynaptic membrane as a result of a decrease in excitatory transmitter release. During the action of hyperpolarizing potential, no change in membrane conductance was observed.

In electroreceptors, the synaptic processes are characterized by a number of peculiarities. For instance, in the absence of stimulation, the sensory cells show a continuous transmitter release, thereby giving rise to tonic impulse activity in the afferent nerve fibers. Stimuli below 1 mV modulated the afferent firing rate. The input-output relation in the electroreceptors appears to be considerably steeper than, for instance, in a squid synapse, and is shifted along the voltage axis toward more negative values of the potential, where the transmitter release occurs in the absence of stimulation. As a result, a small change in the presynaptic potential causes considerable changes in the postsynaptic membrane.

It is worth noting that no miniature potentials characteristic of other receptors of the acousticolateral system have been found in electroreceptors so far. This may possibly be due to a nonquantal release of the transmitter. Moreover, the absence of miniature potentials is probably due to an extensive convergence of receptor cells on afferent nerve fibers, which makes it impossible to record the contribution of individual synapses to the postsynaptic potential.

2.5 The Effect of Ions on Synaptic Processes

Analysis of the influence exerted by different ions upon synaptic processes in the receptors of the acousticolateral system permits better knowledge of some peculiarities of signal transmission from receptor cells to the afferent fibers. Of special interest is the investigation of the effect of divalent ions, which are known to play an important role in different synapses of the peripheral and the central nervous system.

Thus, perfusion of the guinea pig scala tympany with solutions containing an elevated concentration of Mg^{2+} or an addition of Co^{2+} resulted in a

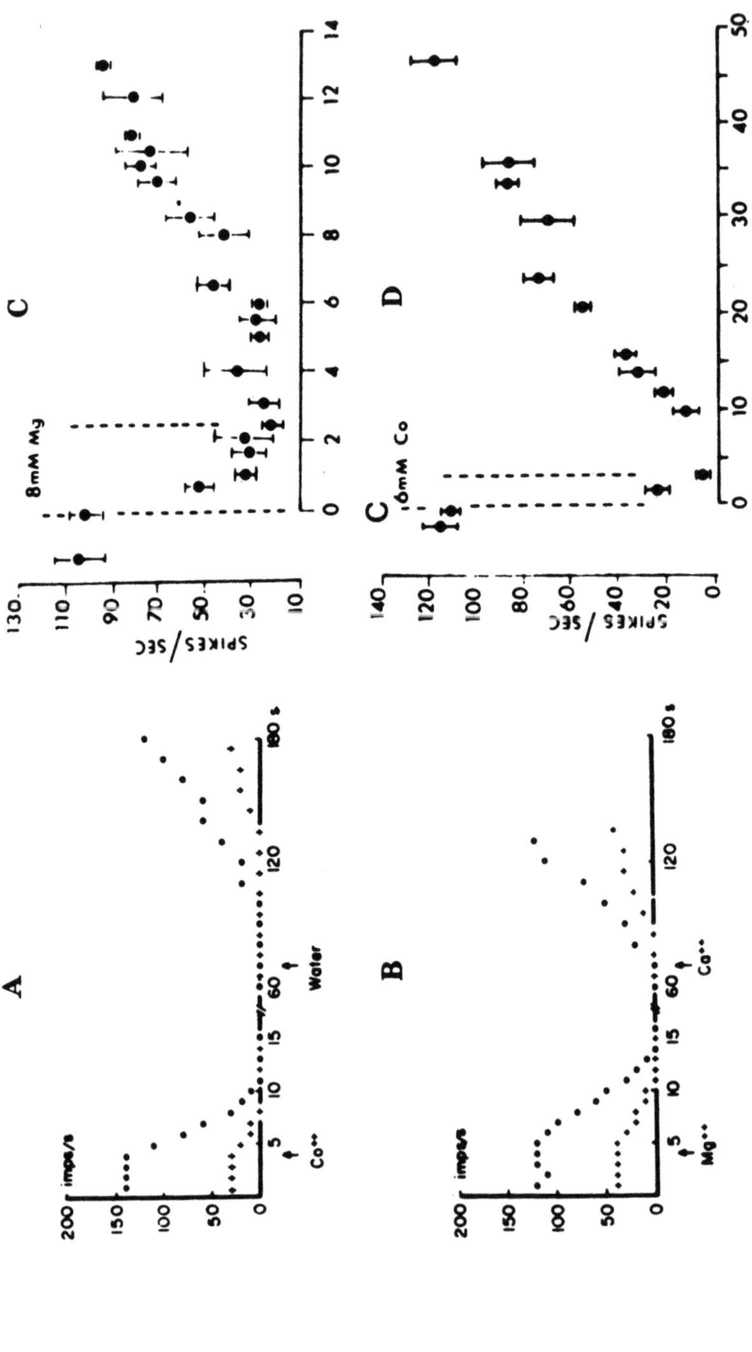

Fig. 50A–D. Effect of divalent cations on the activity of receptors of acousticolateral system. **A,B** Effects of external application of 10 mM Co^{2+} (**A**) and 10 mM Mg^{2+} (**B**) on the resting (●) and evoked (+) nerve discharges in ampullary organs of sturgeon. (Teeter et al. 1980). **C,D** Effects of external application of 8 mM Mg^{2+} (**C**) and 6 mM Co^{2+} (**D**) on resting firing rate of primary auditory neurons of guinea pig. (Robertson and Johnstone 1979)

reversible blocking of resting activity in the single fibers of the cat's auditory nerve (Fig. 50C,D) (Robertson and Johnston 1979). At the same time, raising the calcium ion concentration from 2.5 to 8 mM had no effect on the background activity of the nerve fibers, while a complete removal of these ions caused a decline in their activity. Increasing Mg^{2+} concentration up to 10 mM led to a rise in receptor thresholds, the effect being antagonized by an elevated Ca^{2+} concentration. It is argued that the primary effect of divalent cations is to control of transmitter release by hair cells (Siegel and Relkin 1987). Both resting and evoked activities of the single nerve fibers in the frog labyrinth were also suppressed in a Ca^{2+}-free or Mg^{2+}-rich solution (Fig. 47C) (Rossi et al. 1977; Taglietti et al. 1977; Annoni et al. 1984; Rossi and Martini 1988). In the absence of Ca^{2+}, the postsynaptic activity was restored in a Sr^{2+} (1.8 mM) or Ba^{2+} (0.3 mM) solution. When Ba^{2+} concentration was raised to 5 mM, a dose-dependent increase in the EPSP and spike frequency was observed. In addition, Ba^{2+} can restore EPSPs and afferent firing in Co^{2+} (3 mM) solution. TEA (20 mM), a selective blocker of Ca^{2+}-activated K^+ current elicited a clearcut increase in the EPSP and spike discharges. On the other hand, 4-AP (4–20 mM), a selective blocker of an A-type K^+-current, caused a decrease in sensory activity. The authors believe the drug effect to be localized mainly at the basal face of sensory cells due to their interaction with Ca^{2+}- and K^+-currents.

Synaptic activity in afferent fibers innervating the lateral line of *Xenopus laevis* was suppressed by divalent ions (Russell 1971; Drescher and Drescher 1987a,b, Sewell and Mroz 1987; Guth and Drescher 1990; Sewell 1990). The rank order of potency of divalent ions was $Co^{2+} > Ca^{2+} > Mg^{2+} > Ba^{2+}$ (Guth and Drescher 1990).

The EPSP amplitude of the goldfish auditory nerve decreased in a Mg^{2+}-rich solution and increased in a Ca^{2+}-rich one (Furukawa and Matsuura 1978). The Ca^{2+} effects are more likely to be due to a rise in n and p parameters.

A number of reports are available concerning the influence of divalent ions on synaptic processes in electroreceptors. Thus, removal of Ca^{2+} from the external solution or an increase in the Mg^{2+} concentration blocked the synaptic transmission in the ampullae of Lorenzini of *Raja*; (Steinbach 1974; Nier and Hensel 1973; Clusin and Bennett 1977a; Akoev et al. 1980a; Akoev and Andrianov 1989), and in the tuberous electroreceptors of the weakly electric freshwater mormyrid (Steinbach and Bennett 1971).

Resting and evoked activity of the electroreceptors in the transparent catfish (*Kryptopterus*) disappeared rapidly in a solution with Co^{2+} or Mg^{2+} (Teeter and Bennett 1981) (Fig. 50A,B). Moreover, a high Ca^{2+} (15 mM) solution appeared to raise the threshold for evoked responses without significantly altering the resting discharge. The activity of ampullary electroreceptors of sturgeons was also blocked in a solution with 10 mM Co^{2+} or Mg^{2+} (Teeter et al. 1980). A rise in Mg^{2+} concentration in the external solution produced an increase in the threshold and a decrease in the amplitude

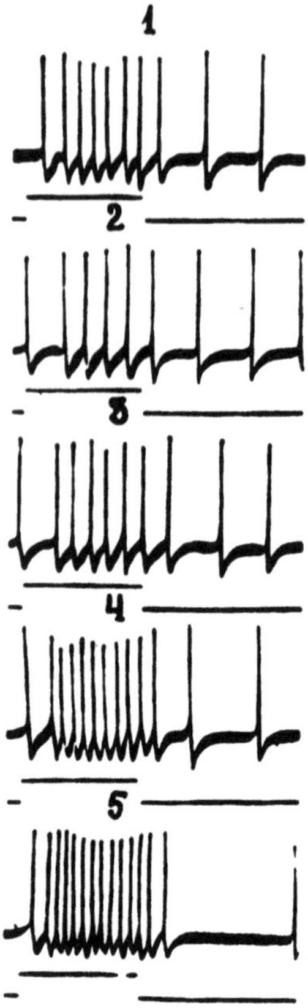

Fig. 51,1–5. Effect of Ca^{2+}-free EGTA (10 mM)-containing solution on responses of the ampullae of Lorenzini to electrical pulses (*lower traces* 0.5 μA, 200 ms). **1** response in normal solution. **2–5** Responses in Ca^{2+}-free solution (5, 20, 50, and 90 min, respectively). (Akoev and Andrianov 1981)

of EPSP of the ampullae of Lorenzini, Ca^{2+} being Mg^{2+} antagonist in these processes (Steinbach 1974). However, at a Mg^{2+} concentration above 10 mM, the changes in the Ca^{2+} concentration became ineffective. The effect produced by La^{2+} was similar to that of Mg^{2+}, only at lower concentrations. A decrease in Ca^{2+} concentration in the solution caused a reduction of the EPSP amplitude, and in the Ca^{2+}-free solution with 5 mM EGTA the synaptic transmission was entirely blocked.

A more detailed study on the Ca^{2+} effect on the ampullae of Lorenzini of *Raja* have revealed that superfusion of the basal membrane with Ca^{2+}-free solution plus 10 mM EGTA affected both evoked and resting activities (Akoev et al. 1980a, 1981). In the Ca^{2+}-free solution, a two-phase change in receptor excitability was observed (Fig. 51). On the one hand, a decrease in Ca^{2+} concentration is known to cause a reduction in the electrochemical gradient for these ions, thus diminishing Ca^{2+} influx during the depolarization of the basal face of receptor cells and, consequently, reducing transmitter release. On the other hand, a decrease in Ca^{2+} concentration reduces the electrostatic screening of the fixed negative surface charges near the Ca^{2+}

channels of the presynaptic membrane, which causes depolarization across the channels, and an increase in the number of open calcium channels. This leads to an increase in Ca^{2+} influx and transmitter release. One may suppose that the phase of decrease in the excitability of the ampullae of Lorenzini in Ca^{2+}-free solution is due to a decrease in the electrochemical gradient for these ions, whereas the phase of excitability increase may be explained by an effective depolarization across the voltage-sensitive Ca^{2+} channels which become sensitive to a small change in the membrane potential.

It is worth noting that changes in the resting activity of the receptors in a Ca^{2+}-free solution occur much later than those in the evoked responses. These data imply that mechanisms which control the resting release of transmitter should differ at least partly from those involved in the evoked release (Bernard 1983; Guth and Drescher 1990; Guth et al. 1991). Indeed, there are data showing that the resting release of transmitters in various synapses depends on the Ca^{2+} concentration in the cytosol, whereas the mechanism of the evoked release depends on the external Ca^{2+} concentration. These two processes are characterized by different temperature sensitivity (Barrett et al. 1978). Drescher and Drescher (1987a) proposed that intracellular storage sites for Ca^{2+} (mitochondria, synaptic vesicles, etc.) near the presynaptic apparatus may be responsible for spontaneous transmitter release in the *Xenopus* lateral line. Hence, since the hair cells are of epithelial origin, it seems quite possible that spontaneous transmitter release, as in other non-neural secretory cells, may be regulated by intracellular Ca^{2+}. Moreover, the resting discharge seems to be related to the trophic function of the afferent nerve (Peters et al. 1988).

The available data on the role of divalent ions in synaptic transmission in the receptors of the acousticolateral system provide evidence that Ca^{2+} is essential for transmitter release processes. The existence of calcium channels in the basal face of hair cells was demonstrated in the vestibular receptors of the chick (Ohmori 1984a,b). The Ca^{2+} influx through these channels, on the one hand, triggers the transmitter release and, on the other hand, activates the K^+ channels. Indeed, the experiments on the ampullary electroreceptors in the marine catfish *Plotosus* demonstrated the existence of oscillatory responses of receptor cells blocked by superfusion of the basal membrane with TEA and 4-AP (Sugawara and Obara 1984a). These responses were also blocked by Co^{2+} and Cd^{2+}. The authors believe that the basal membrane of electroreceptor cells have Ca^{2+}-activated K^+ channels primarily responsible for the initial phase of receptor adaptation. The presence of Ca^{2+}-activated K^+ conductance may also be important for the stable operation of the highly sensitive receptors.

The apical part of the hair cells of the cochlea and the vestibular system (Murray and Potts 1961; Sellick and Johnstone 1975; Flock 1977), the lateral-line canal organ (Ilyinsky and Krasnikova 1971; Liddicoat and Roberts 1971; Russell and Sellick 1976), and ampullary electroreceptors (Ilyinsky and Krasnikova 1971, 1974; Murray and Potts 1961; Okitsu et al. 1978) are well known to be immersed in a fluid with a high K^+ concentration. The endolymph of the mammalian cochlea contains a high K^+ concentration (150 mM) and a small amount of Na^+ (1 mM). In free-standing neuromasts of *Xenopus*, the cupular microenvironment is rich in K^+ (24–100 mM) and Cl^- (35–70 mM) with a positive endocupular potential (15–50 mV) (Russell and Sellick 1976). Measurements of endocupular Ca^{2+} were found to be in the range of 2–30 µM (McGlone et al. 1979).

In the electroreceptors of *Plotosus*, the ampullary jelly contains 63 mM K^+ (Okitsu et al. 1978) whereas skate jelly contains only 12.5 mM K^+ (Murray and Potts 1961). Analysis of *Plotosus* ampullary electroreceptors revealed that the DC potential in the lumen was near 0 mV in situ and became negative by −13 to −19 mV when the shunt was removed by exposing the duct to air (Sugawara 1989b). Thus, the DC potential provides and outward current to the basal face of the sensory cells which activates Ca^{2+} channels and evokes the continuous release of the afferent transmitter. Under voltage clamp in a 4 mM K^+ solution, the outward current was about 10 nA and was suppressed by ouabain. This current was due to an electrogenic Na^+-K^+ pump in the basal face of the sensory epithelium (Sugawara 1989b).

Davis (1965, 1983) suggested that shearing displacement of the stereocilia of hair cells in the cochlea causes a change in the ohmic resistance across the top of the hair cells, thereby modulating the current flow through the apical membrane and transmitter release. This hypothesis is believed to apply to other hair cells of the acousticolateral system in general (Flock 1971). Evidence has been recently presented to support the "resistance microphone" theory of Davies. Potassium ions were shown to carry a major part of the receptor current in hair cells in the frog sacculus (Corey and Hudspeth 1979a) and semicircular canal (Valli et al. 1977), the turtle cochlea (Crawford and Fettiplace 1981), and the guinea-pig cochlea (Russell 1983). The sound-evoked increase in the K^+ level in the tunel of the guinea-pig organ of Corti by 1–3 mM (Johnstone et al. 1989) and by 0.2 mM in the perilymph of semicircular canal of the frog (Botta et al. 1990) have been recently demonstrated. The reversal potential of the receptor current occurred at a potential close to the endocochlear potential in these preparations. Moreover, the amplitude of the DC component of the RP was linearly related to the resistance change associated with it (Russell 1983; Russell and Sellick 1978).

On the other hand, some data are not consistent with Davies' theory (Leng 1980). Russell and Sellick (1978) observed changes in the electrical resistance of hair cells during sound stimulation. However, their results imply that the current flow through the hair cell was not modulated by these resistance changes but remained constant over a wide range of sound frequency and intensity. Moreover, the sensitivity of inner hair cells in the mouse cochlea in vitro was similar to that of hair cells in vivo (Russell et al. 1986a), and the asymmetrical distribution of K^+ ions was not essential to the transduction process (Ohmori 1984a). Thus, in the absence of endocochlear potential, hair cells in vitro retain a high sensitivity, 0.4 mV per nm

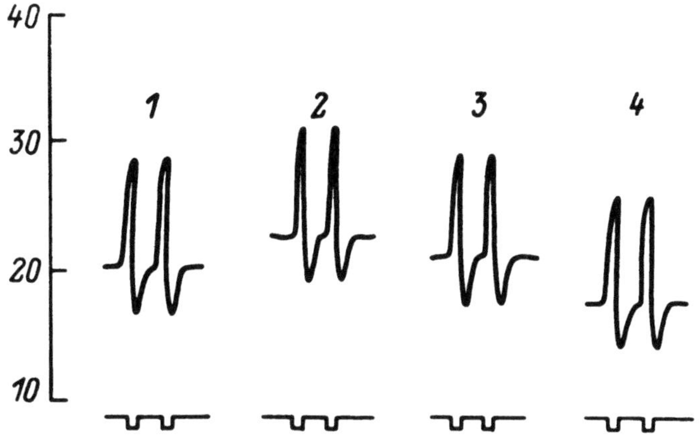

Fig. 52,1–4. The effect of high-K^+ solution (10 mM) on the responses of the ampullae of Lorenzini to electric stimuli. **1** In normal solution. **2–4** In high-K^+ solution (5, 7, and 10 min, respectively). Stimulus duration (*lower trace*) 1 s, intensity 0.4 μA. *Abscissa* imp/s. (Akoev et al. 1984)

displacement of their stereocilia. The rise in the concentration of K^+ ions in the basal face increased the firing rate in the hair cells of the organ of Corti (Leng and Comis 1979a,b), the frog semicircular canal (Rossi and Martini 1988; Valli et al. 1977, 1988), the lateral line of fish (Hashimoto and Katsuki 1972), and the ampullae of Lorenzini (Akoev et al. 1984a,b). Spontaneous synaptic potentials of frog vestibular afferent fibers increased in frequency but with little change in amplitude in 3 mM K^+ solution (Annoni et al. 1984).

The most interesting result is that the threshold hair cells of the guinea-pig cochlea and the ampullae of Lorenzini produced no changes in a K^+-high solution (Leng and Comis 1979a; Akoev et al. 1984a,b) (Fig. 52). These data do not seem to be consistent with Davies' theory. Thus, a change in the hair cell membrane potential evoked by a high K^+ concentration did not have the same manipulations that change the endocochlear potential. This implies that a change in endocochlear potential is not directly responsible for the altered hair cell responses to mechanical stimuli.

According to Davies' theory, there is no practical difference in the mechanism of spontaneous and evoked releases of neurotransmitter. However, at present, the spontaneous and evoked releases of the transmitter are believed to be based on different mechanisms (Bernard 1983; Peters et al. 1988). Indeed, Zucca et al. (1982) demonstrated that in Ca^{2+}-free solution the frog ampullary receptors did not respond to mechanical stimuli. However, despite the suppression of mechanosensitivity, these receptors were still able to maintain resting discharges in afferent fibers.

The role of high K^+ concentration in the media surrounding the hair cells is still obscure. K^+ may be essential for the specific mechanical properties of the tectorial membrane, or activation of transduction channels. Sensitivity of stretch-activated channels in chick muscle (Guharay and Sachs 1984) and that of the Pacinian corpuscle (Akoev and Elman 1974) were greatly enhanced by a high extracellular K^+ level.

Lawrence (1975) believed that the potential of the inner sulcus was zero, and thus open communication between this space and scala media should have been absent. This does not mean that the ion concentration within the inner sulcus differs from that of the endolymph. Moreover, other parts of the mammalian or nonmammalian inner ear are characterized by the lack of a large positive potential. The inner sulcus of the cochlea of chinchillas and the subtectorial space above the hair cells have recently been shown to be separated by a band of glycoconjugate (Santi and Andersson 1987). Thus, the apical face of hair cells does not come into contact with the endolymph. Interestingly, the K^+ concentration in the fluid of the inner sulcus of the guinea pig is similar to that in the perilymph, i.e., 2–4 mM (Runhaar and Manley 1987). This finding is in conflict with that of Flock (1977), who showed a high content of K^+ in the inner sulcus in the inner ear of the guinea pig. The role of K^+ and endocochlear potential in the mammalian cochlear remains to be established.

The data that are presented in this chapter testify to considerable progress achieved in the study of the properties of hair cells in the acousticolateral system, and the role played by different ions in the generation of the RP, as well as in the analysis of synaptic processes. The first intracellular recordings of potentials were performed in the early 1970s. To the present time, such investigations have been carried out on different hair cells. The membrane potential of hair cells has been established to be −40 to −70 mV, i.e., these cells do not differ practically in membrane potential level from other types of nerve cells. In response to adequate stimulation, hair cells generate a depolarizing RP of 10–25 mV and a hyperpolarizing RP with an amplitude two to three times smaller than that of the depolarizing one.

It should be noted that RP generation is due to the displacement of stereocilia, whereas kinocilia are not actively involved in transduction processes. Only several mechanically activated channels have been found to be located in the distal part of stereocilia. Recently, important characteristics have been identified in isolated hair cells; a mechanical displacement of their stereocilia induced receptor current of 20 pA. The nonlinearity of hair cell current-voltage characteristics is associated with the delayed and anomalous rectification of sensory cells. In its turn, the mechanoreceptor current depends linearly upon the level of the membrane potential and the magnitude of the mechanical stimuli.

The mechanotransduction channels of hair cells are nonspecific and are permeable to mono- and divalent cations. The endolymph in vertebrates is characterized by an elevated content of potassium ions, and it is by these ions that the transduction current is carried. It is important to note that the transduction channels are permeable to calcium ions as well but, according to many authors, these are involved to a greater degree in channel activation processes.

It has been established on isolated hair cells that during depolarization several types of ion currents can be observed: a voltage-dependent Ca^{2+} current activated in part at the resting potential and involved in the process of transmitter release; a Ca^{2+}-activated potassium current, and a fast potassium current. Depolarization (receptor potential) of hair cells evoked by the transduction current activates voltage-sensitive Ca^{2+} channels. Ca^{2+}, in turn, activates the K^+ channels, leading to membrane repolarization.

During the depolarization of electroreceptor cells the voltage-sensitive Ca^{2+} current of the apical and basal membranes, the Ca^{2+}-activated potassium current, and the voltage-dependent potassium current are observed.

In the afferent fibers of the acousticolateral system, spontaneous and evoked EPSPs have been recorded. The spontaneous EPSP amplitudes are 0.5–5 mV, with 3–6 ms duration. Under the action of inhibitory stimuli, a hyperpolarizing generator potential arises that is due to the repolarization of the postsynaptic membrane as a result of a decreased release of the excitatory transmitter. Hair cells are assumed to contain many transmitter release sites with different sensitivity to membrane depolarization. In electroreceptor cells no miniature postsynaptic potentials characteristic of other receptors in the acousticolateral system have been found so far; this may possibly be due to the nonquantal transmitter release.

The data available on the role of divalent ions in synaptic transmission in the receptors of the acousticolateral system provide evidence that Ca^{2+} is essential to transmitter release processes.

Chapter 3

The Mode of Action of Amino Acid-Using Synapses

The use of the physicochemical, electrophysiological, and pharmacological techniques of studying synaptic transmission permits the identification of neurotransmitter and also the investigation of the peculiar mechanisms of its synthesis, release, interaction with postsynaptic membranes, and inactivation. When considering a chemical as a candidate for a neurotransmitter in the synapse, a number of criteria need to be satisfied in order to establish it as the endogeneous neurotransmitter (Orrego 1979; Guth and Melamed 1982; Klinke 1986; Bledsoe 1986; Bledsoe et al. 1988; Shinozaki 1988). Among these are the following: (1) The presence of the transmitter candidate and the possibility of its synthesis in the presynaptic elements; the presence of the necessary enzymes and the appropriate precursors. (2) Calcium-dependent release of the transmitter candidate. (3) An increase in the release of this substance during stimulation by physiological means. (4) The presence of the mechanism of the enzymatic inactivation of transmitter candidate, or its presynaptic uptake. (5) Identity of the postsynaptic effects of the transmitter candidate and of the endogenous transmitter. (6) Identity of the influence of pharmacological substances on the effects of the transmitter candidate and the endogenous transmitter. (7) The existence of postsynaptic receptors specific for the transmitter candidate.

At present, a great number of chemical substances have been found to satisfy the criteria for neurotransmitter candidates and they are probably capable of performing the neurotransmitter function in the synapses of the peripheral and the central nervous system. Among them, the following substances have been advanced: catecholamines, acetylcholine (ACh), γ-aminobutyric acid (GABA), a GABA-like compound, serotonin, dopamine, the auditory nerve-activating substance (ANAS), substance P, and excitatory amino acids (L-glutamate and L-aspartate) (L-GLU and L-ASP). The latter appear to be of greatest interest because the evidence best supports the hypothesis that the transmitter in many synapses of the peripheral and central nervous systems in vertebrates and invertebrates is an excitatory amino acid (Werman 1966; Gerschenfeld 1973; Krnjevic 1974; Mandelshtam 1983; Fonnum 1984; Foster and Fagg 1984; Roberts 1984; Klinke 1986; Bledsoe 1986; Takeuchi 1987; Piotrovsky 1987; Bledsoe et al. 1988; Shinozaki 1988; Akoev and Andrianov 1988, 1989). For this reason the following discussion will be mainly devoted to afferent transduction and the transmitter role of the excitatory amino acids.

3.1 Release of Amino Acids

One of the primary criteria for identifying the transmitter function of a substance is the demonstration of its specific release from presynaptic endings upon the stimulation of the end-organ (Werman 1966; Fagg and Lane 1979; Orrego 1979). Such experiments are usually performed on brain tissue slices, isolated synaptosomes, preparations of isolated receptors or, in vivo, on the structures of the central nervous system. For this purpose, preloading the preparation with a radiolabeled amino acid and subsequent monitoring of the background and evoked efflux of radioactivity are usually employed. However, the demonstration of the evoked release of endogenous ligands by natural stimuli is most desirable for this type of investigation. The efflux of endogeneous glutamate is usually difficult to demonstrate, therefore, the release experiments are often performed using the labeled amino acid taken up into the cells, and the K^+-rich solution or electrical current as the stimulus. Interpretation of such data may often be difficult due to the absence of data concerning the concrete mechanisms of putative transmitter release from the terminals in vitro and the identity of the mechanism of release in vitro to the processes in vivo.

The method of investigating the release of putative amino acid transmitters on brain slice preparations found wide application. For instance, using this technique, it was possible to show that stimulation of the frontal cortex resulted in a significant increase in the release of labeled L-GLU from the neostriatum in rats (Godukhin et al. 1979). This demonstrated the glutamatergic nature of the rat corticostriatal fibers. By electric stimulation, a considerable increase in the release of radiolabeled L-GLU and GABA from the striatal and cortical slices of the rat brain was observed (Katz et al. 1969; Hammerstad and Cutler 1972), whereas a release of glycine, GABA, and L-GLU was observed from spinal cord slices of the same animals (Hammerstad et al. 1971). In the same preparations, a release of another physiologically active amino acid, taurine, was noted (Katz et al. 1969; Hammerstad et al. 1971). It is important that, when applying a similar method to liver and kidney slices, no GABA or L-GLU release was observed, suggesting that evoked amino acid release was a property inherent only to neural tissue.

The method of studying radiolabeled L-GLU release was also successfully used to study the mechanisms of synaptic transmission in the neuromuscular synapses of insects (Usherwood et al. 1968), crayfish (Krawitz et al. 1970; Wang and Boyarsky 1979), and other animals. However, the significance of the uptake and release of exogenous labeled substance and also its intracellular distribution is not clear from these experiments. Some authors (Reubi 1980) consider that more solid data can be obtained not by saturating the preparation with labeled L-GLU, but by studying the release of L-GLU synthesized from its labeled precursor (glucose or glutamate).

An alternative means for the stimulation of neurotransmitter release is to raise the extracellular potassium ion concentration (Hillman and McIlwain 1961; Fagg and Lane 1979). An increase in amino acid efflux was observed only after a considerable rise in K^+ concentration (usually within 40 to 100 mM), whereas solutions with lower concentrations of K^+ (25–30 mM) did not produce any change in the amount of radiolabeled amino acids from striatal slices (Katz et al. 1969). A detailed analysis of the process of amino acid release suggests the existence of some dependence between K^+ concentration in the medium and the composition of the fluid released from the cells (Mulder and Snyder 1974). It was observed that raising K^+ concentration higher than 14 mM caused an increase in the efflux of glycine, in excess of 28 mM an increase in glutamate release was shown, and higher K^+ levels were accompanied by a considerable rise in GABA release.

The development of effective mass fragmentography using a gas chromatograph-mass spectrometer (Miledi et al. 1977) was a very useful method to determine the concentration of glutamate during control rest period and natural stimulation. It was shown that in the crayfish junction the background efflux of glutamate was 20 pM/10 min (Kawagoe et al. 1981) (Fig. 53). During nerve stimulation with electric stimuli at a frequency of 10 to 20 Hz, the increase in the net release of glutamate above the background was 11.9 and 21.1 pM/100 µl, respectively. It is curious that perfusion with streptomycine, decreasing the amplitude of excitatory postsynaptic potentials (EPSP), did not affect the net release of glutamate during stimulation, which is a convincing proof of glutamate release from the presynaptic terminals. This conclusion was also supported by the dependence of the amount of glutamate released upon the external concentration of calcium ions. At the same time, nerve stimulation had little effect on the release of exogeneous glutamate. It was suggested that endogenous glutamate may be stored in synaptic vesicles while exogeneous glutamate may be taken up into compartments separate from the endogeneous transmission pool, probably into glia (Kawagoe et al. 1982, 1984). Aspartate was also released upon dorsal root stimulation, but the release was several times smaller than that of glutamate and statistically not significant (Kawagoe et al. 1985, 1986).

Studying the processes of neurotransmitter release, it should be borne in mind that the fluid sampled for analysis may contain not the neurotransmitter itself but its derivatives, due to degradation of the neurotransmitter in the period between its release and sampling. Additionally, transmitters and substances accompanying them are released by means of the mechanism of exocytosis, which requires the influx of Ca^{2+} into presynaptic elements. However, the stimulation of the end-organs may also activate other processes unrelated to exocytosis, which may also cause a release of nontransmitter

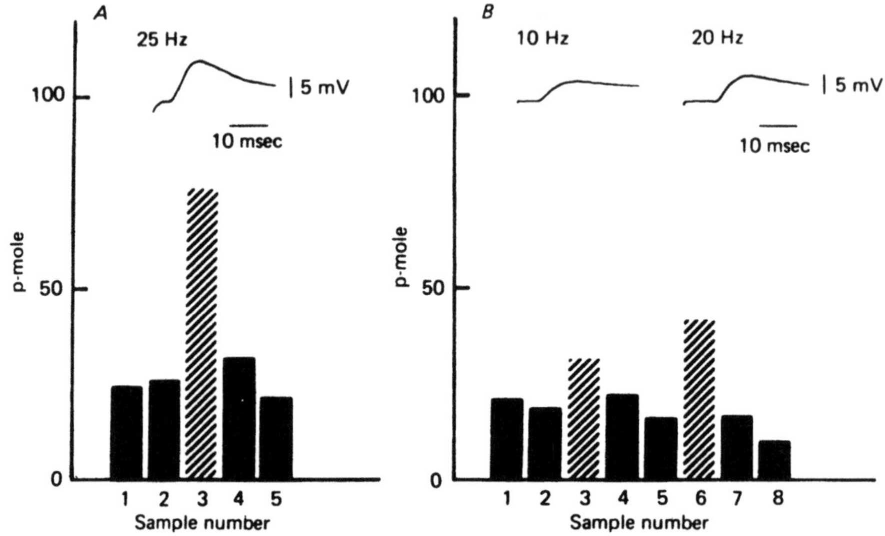

Fig. 53A,B. Examples demonstrating the release of glutamate from the crayfish neuromuscular junction in response to nerve stimulation. Each *column* represents glutamate content in 100 µl solution for a sampling period of 10 min. A Nerve stimulation was applied at 25 Hz for 7 min, and the glutamate content in this solution is shown in the *hatched column*. **B** Stimulation was applied at 10 Hz (number *3*) and 20 Hz (number *6*) for 7 min. Sample records of EPSP are shown in *inset*. (Kawagoe et al. 1981)

substances, i.e., modulators. Indeed, during electric stimulation of the reticular formation, an increase in the concentration of transmitter amino acids including alanine, GABA, GLU, and glycine was observed. At the same time, there was a release of an inactive amino acid, glutamine (Blagoeva et al. 1972). This process may be explained by the accumulation of intracellular Na^+ during continuous stimulation, and its activation of the system of transport carriers for some group of structurally related amino acids (Orrego 1979). In parallel, physiological stimulation also causes an increase in extracellular K^+ which in turn increases neuronal depolarization and Na^+ influx. It also depolarizes glial cells and stimulates the mechanisms of amino acid release.

3.2 The Mechanisms of Neurotransmitter Release

Studies on neuromuscular junction (del Castillo and Katz 1954; Katz 1962; Katz and Miledi 1967), as well as on a number of other synaptic structures, suggest that depolarization of the nerve ending membrane is a necessary but insufficent condition for the release of neurotransmitter. Ca^{2+} deficiency in the medium considerably inhibited the process of transmitter release without essentially affecting the sensitivity of the postsynaptic membrane to the transmitter. These findings were supported by the observation that the application of a low Ca^{2+} solution produced a significant decrease in both the frequency of miniature potentials and the amplitude of the end-plate potentials in the neuromuscular synapse. Studies on the crayfish giant synapse have shown that a decrease in postsynaptic potentials with a reduction of the outer Ca^{2+} concentration was not accompanied by a decay of the amplitude of spikes in presynaptic endings (Takeuchi and Takeuchi 1962). The role of Ca^{2+} in the process of neurotransmitter release was also demonstrated on motoneurons (Shapovalov 1966) and many other objects (Orrego 1979). Thus, the current classical studies have positively demonstrated the important role of Ca^{2+} in transmitter secretion.

At present, substantial differences have been revealed between the processes of transmitter release in different synapses. It was found that, contrary to the absolute calcium requirements of dopamine, ACh, noradrenaline, and 5-hydroxytryptamine release, a similar process for amino acid neurotransmitters is characterized by a more complicated dependence on the Ca^{2+} concentration in the extracellular medium (Katz et al. 1969; Hammerstad et al. 1971; Redburn et al. 1976; Orrego et al. 1976; Levi et al. 1980; Orrego 1979; Sanchez-Prieto et al. 1987; Nicholls et al. 1987). It turned out that in many cases the process of amino acid release depended on the outer concentration of Ca^{2+} if depolarization was evoked by potassium ions; if it was caused by electric stimulation or veratridine, its release could also often take place in a calcium-free medium. Moreover, it was found that taurine release was even enhanced in a calcium-free medium. At a normal concentration of Ca^{2+}, the process of amino acid release does not depend on the kind of depolarizing action. It is curious that replacing Ca^{2+} by Co^{2+} suppressed amino acid release during K^+ depolarization, yet this process not only persisted but was even increased during electric stimulation (Godukhin et al. 1983). The data obtained concerning the type of depolarizing action upon the process of transmitter release by no means contradict the Ca^{2+} hypothesis but serve to support it. Indeed, it may be supposed that K^+ depolarization causes a Ca^{2+} inward current from the extracellular medium into the cell which triggers a transmitter release. This process is blocked by cobalt ions. During electric stimulation, calcium ions are also necessary for neurotransmitter release; but a rise in their concentration within the cell seems to be due to Ca^{2+} efflux from intracellular organelles. That is why this process does not depend on the external Ca^{2+}-concentration. However, in both cases, the process of transmitter release depends upon free calcium content inside the cell and to a lesser degree it is determined by Ca^{2+} concentration in the extracellular medium. Characteristics, conditions, and possible functional significance of external Ca^{2+}-independent neurotransmitter release are extensively reviewed elsewhere (Adam-Vizi 1992).

Since electric field stimulation often caused calcium-independent amino acid release whereas potassium depolarization was, on the contrary, connected with calcium-dependent release of a number of putative neurotransmitters, it has been suggested that potassium depolarization constituted a more physiological technique to study the release of endogenous amino acids and other primary amines (Mulder and Snyder 1974; Bledsoe et al. 1989a). It should be noted that this point of view is not shared by all authors. On the contrary, it is suggested (Orrego et al. 1976) that K^+ induces a nonspecific release of amino acids from the cytoplasmic pool of the neurotransmitter. Different mechanisms of behavior of exogenous and endogenous amino acids upon nerve stimulation and high K^+ depolarization also cannot be excluded. Only additional experiments seem to be able to solve this problem. In particular, the role of glial and other nonsynaptic elements in the process of neurotransmitter release should be clarified, as well as the significance of multiple sites of its storage.

The intracellular concentration of calcium ions plays a decisive role in the mechanism of neurotransmitter release. The cytoplasm of many cells is known to contain a very low concentration of Ca^{2+} (10^{-8}–10^{-7} M), and a whole number of mechanisms is directed at maintaining this concentration (Perney et al. 1984). There exist experimental data obtained on tissue culture showing that in the course of continuous electric stimulation a reduction of inward Ca^{2+} current and an accompanying decrease of transmitter release take place (Jia and Nelson 1986). The inability to perform the synaptic function during afferent input activation in a number of objects appears to be directly connected with intracellular calcium accumulation, which considerably affects the calcium current magnitude and the amount of transmitter released. It is interesting that in such preparations, synaptic boutons contain relatively fewer mitochondria than those in the spinal cord, where the processes of synaptic inactivation are less developed. Therefore, maintaining a low intracellular concentration of Ca^{2+} is one of the main functions of the cell, and the only mechanism in this process seems to be a transfer of free intracellular calcium into the endoplasmatic reticulum and mitochondria.

There is a traditional conception that the transmitter is released from the presynaptic endings not only as a result of stimulation but also spontaneously. This produces excitatory postsynaptic potentials which are referred to as miniature potentials. The appearance of miniature potentials is conditioned by the influence of the transmitter quantae upon the postsynaptic membrane. The amplitude of extracellular potentials recorded is very stable, which suggests that the transmitter is released as approximately equal packets, which are located in the vesicles of the presynaptic endings. At present, the quantal theory of transmitter release is generally accepted, yet there is evidence that some amino acids can also be released nonquantally. Therefore, of great importance are the studies in which the amino acid content was investigated directly in synaptic vesicles and synaptosomes (De Belleroche and Bradford 1973). A special chemical analysis of synaptic vesicles released from the neurons of the brain cortex showed that glutamate, aspartate, GABA, glutamine, serine, glycine, alanine, and taurine are contained only in trace quantities. Amino acid content in the vesicular fraction was only 0.4–1.3% of the intracellular pool of amino acids. In the vesicular set of amino acids, glutamate and taurine showed the highest concentration. This suggested that in some objects amino acid release may be not only quantal but may also occur by means of membrane carriers which transport transmitter molecules across the presynaptic membrane. In this case, the main source of physiologically active amino acids released from synaptosomes is a nonvesicular but cytoplasmatic component of the neurotransmitter. The principal possibility of nonvesicular transmitter release was demonstrated for acetylcholine in the frog neuromuscular synapse and the skate electric organ (Kelly et al. 1979).

The absence of amino acids in the fraction of synaptosomes and synaptic vesicles in the above studies may be partly explained by the fact that the transmitter amino acids are bound to synaptic vesicles differently than ACh (Fonnum 1984). This required new techniques for the isolation of proper synaptosomal and vesicular pools of glutamate. Indeed, recently, in many presumed glutamatergic pathways of CNS, fast retrograde D-ASP transport was noted (Streit 1980), an ATP-dependent uptake of glutamate by synaptic vesicles isolated from bovine cortex was registered (Naito and Ueda 1983, 1985), and a glutamate concentration of up to 50 nM per mg of vesicular protein was found (Disbrow and Ruth 1985). At the morphological level, data on intravesicular localization of 4-aminobutirate and glutamate have been obtained (Storm-Mathiesen et al. 1983). All these data are considered in connection with the existence of glutamate-accumulating synaptic vesicles.

3.3 A Transmitter Inactivation Mechanism

By definition, one of the primary criteria for the identification of a chemical transmitter substance is the presence of the transmitter candidate and the possibility of its synthesis in the presynaptic elements. In a number of cases this approach was fruitful, which made it possible to suggest a number of substances including GABA, ACh, and catecholamines as potential candidates. These substances or their enzyme systems were found on using histochemical, autoradiographic, and immunohistochemical techniques on brain tissues (Vinnikov and Titova 1961; Fuxe 1965). The identification of amino acids as neurotransmitters in different regions of the nervous system is more difficult, as amino acids play an extensive metabolic role in the organism. For example, L-GLU is a structural unit in protein synthesis; it is connected with the detoxification of ammonia in the brain, the transamination reaction, the synthesis of proteins, peptides, and amino acids including glutamine, GABA, glutathione, etc. (Meister 1979; Fonnum 1984). The result is that in brain tissues, about 30% of the total content of free amino acids belongs to L-GLU (Homma 1983). Therefore, it is reasonable that even the demonstration of enzymes involved in L-GLU metabolism (such as glutamatehydrogenase, glutamatedecarboxylase, glutaminotransferase) could not serve as evidence of the glutamatergic type of transmission.

Evidence for a chemical as the afferent transmitter in the peripheral and central nervous system is the presence of a mechanism that can either remove the transmitter candidate from the synaptic cleft or inactivate its physiological influence on the postsynaptic receptors. This mechanism may include: (1) enzymatic degradation and (2) transport processes. The first mechanism was found to exist in the cholinergic synapses of the neuromuscular junction (Hildebrand et al. 1974).

The most important point in the study of the mechanisms of amino acid pathways in the brain was the proof that enzymatic amino acid inactivation in the synaptic cleft is not significant, and highly specific transport, or uptake, into nerve endings and glial cells may represent a more widespread means of transmitter inactivation. At present, two kinetically different mechanisms of amino acid uptake in brain synaptosomes are known to exist. The neuronal mechanisms of uptake are usually very specific for certain substances, therefore they are referred to as high affinity systems. They are characterized by a rather low Michaelis constant (K_m with values less than 100 μM), as opposed to low affinity less specific uptake systems (K_m = 0.1–1 mM). High affinity uptake systems seem to be confined to amino acids that are considered as putative neurotransmitters. They allow cells to accumulate some compounds against the concentration gradient. These systems are usually sodium-dependent and utilize carrier systems operating at the cell membrane. They may function to clear the synaptic cleft of the transmitter, thus stopping its action, or they may clear the cleft of transmitter breakdown products following inactivation by some other mechanisms. The dependence of the uptake of labeled amino acids upon the influence of the inhibitors of metabolic processes such as ouabain and 2,4-dinitrophenol, and also upon low temperature, are known (Iversen and Neal 1968; Neal 1971; Logan and Snyder 1972). At present, high affinity systems have been described in the brain for ASP, dopamine, GABA, glycine, GLU, serotonin, proline, and noradrenaline. On the contrary, low affinity systems seem to be associated with the transport of substances that are not directly involved in the performance of the synaptic function. For example, for glycine, which seems to be an inhibitory transmitter in the spinal cord, both high and low affinity transport systems of uptake have been noted, whereas in the cerebral cortex, where its involvement in synaptic transmission is hardly probable, only a low affinity system of uptake was found (Watkins 1978). These data suggest that a high affinity transport system may be associated with a neurotransmitter function of amino acids, and the low affinity systems possibly subserve the general metabolic function.

At the first stage of the investigation of transmitter uptake, the high affinity system was associated with the Na^+-dependent transport of substances by presynaptic neuronal membrane (Iversen and Neal 1968; Logan and

Snyder 1972; Johnston 1981; Fonnum 1984; Schank et al. 1987). Recently, however, great attention has been paid to the function of transport systems in glial elements. Glial cells were also shown to be characterized by a Na^+-dependent high and low affinity uptake of amino acids (Hertz 1979). Not only transmitter substances, but also other amino acids are subject to uptake (Peterson and Raghupathy 1978). The existence of such data suggests that both uptake systems take part in transmitter inactivation but show different activity, depending on the experimental conditions. There is an opinion that high affinity systems contribute less to the inactivation of exogenous amino acids in a relatively intact tissue than in a fragmental one. High affinity systems seem to be less accessible to exogenous amino acids than low affinity ones in the intact tissue (Watkins 1978). The result is that under experimental conditions glial cells serve as the main site of exogenous amino acid uptake. This does not at all mean that the endogenous amino acid released from presynaptic endings is also taken up by glial cells. In a number of experiments on preparations of synaptosomes and glial cells of cerebral cortex, it was shown that neuronal amino acid uptake prevailed considerably over the glial uptake of L-GLU and L-ASP, and constituted about 70–80% of the synaptosomal (Weiter et al. 1979; Storm-Mathiesen and Iversen 1979).

On the other hand, there is evidence (Storm-Mathiesen 1977, 1981) that degeneration of synaptic terminals in some areas of the hippocampus, as a result of axotomy, caused a considerable reduction in L-[^{3}H]-GLU uptake by the high affinity system. It is very probable that the high affinity uptake system in synaptic endings exists both for stopping the neurotransmitter action and for accumulation of the transmitter in the absence of stimulation, whereas glial amino acid uptake serves to reduce the diffusion of the neurotransmitter from the synaptic cleft during synaptic activity, to convert the neurotransmitter into the nonactive precursor, and to return it subsequently to the neurons.

Recently, the existence of high affinity uptake systems, which are even more specific than those for amino acid systems, has been proved. Thus, the intermediate component of the Krebs cycle, 2-oxoglutarate, was found to be taken up by glutamatergic neurons more actively than GLU itself (Carter et al. 1986). After a bilateral transection of the glutamatergic tract, high affinity uptake of [^{14}C]-GLU in rat striatal synaptosomes was reduced by 33%, GABA uptake did not change, whereas that of 2-oxo-[^{14}C]-glutarate was reduced by 67%. A relatively small reduction in GLU uptake after the transsection of the glutamatergic tract seems to be partly explained by the existence of the mechanism of GLU uptake by glial cells. The presence of a high affinity uptake system of 2-oxo-glutarate makes it a convenient and selective marker in detecting glutamatergic neurons.

In studying the uptake of amino acids, an important role is played by the inhibitors of this process. They are used for further characterization of the synaptic transmission when sampling the fluid analyzed for the content of amino acids during synaptic activation; they are efficient in clarifying the role of certain transmitter candidates whose pharmacological activity is enhanced in the presence of uptake inhibitors, in contrast to compounds that are poor substrates for the uptake system (e.g., kainate). A wide range of inhibitors is also used for the investigation of the structural specificity of the uptake systems. However, interpretation of such data presents serious difficulties, since the inhibitors of uptake systems available at present are either substrates for uptake themselves or are agonists of excitatory amino acids. The best-known inhibitors of uptake are Tris-3-hydroxyaspartate and dihydrokainate (Johnston et al. 1974, 1979, 1980; Balcar et al. 1977; Lodge et al. 1980). Both substances increased the responses of spinal neurons to L-GLU, L-ASP, and quisqualate (Q) application, but did not affect responses evoked by synaptic activation. D- and L-isomeres of Treo-3-hydroxyaspartate seem to be substrates for the uptake systems and also amino acid agonists. Dihydrokainate is not subject to uptake and is a weak agonist. It caused a slow depolarization of spinal motoneurons that seems to be induced by accumulation of endogenous transmitter released (Watkins and Evans 1981).

3.4 Characteristics of Amino Acid Binding Sites

An important role in understanding the mechanisms of synaptic transmission is played by the processes of neurotransmitter interaction with postsynaptic receptors. Recently, ligand binding techniques have been developed as a direct method of investigating amino acid receptors. Ligand binding techniques make it possible to make direct measurement of ligand-receptor interaction, determine the factors influencing the process of transmitter binding with synaptosomes, localize the site of the drug-receptor interaction, determine the affinity of various ligands to the receptor site, localize receptors in different brain areas and subcellular fractions, monitor the isolation of the receptor molecule, and determine its biochemical parameters.

The method of ligand binding involves saturation of the preparation that contains the receptor with radioactive ligand, determination of the radioactive ligand and of the radioactivity bound to the tissue, and subtraction of appropriate blanks. By means of the ligand binding method, such receptor characteristics can be determined as the equilibrium dissociation constant (K_d) and the maximum number of binding sites (B_{max}). For a detailed description of the technique and comprehensive data on the binding of amino acids see Scatchard (1949), Yamamura et al. (1978), Foster and Fagg (1984), Dambinova (1989), and Young and Fagg (1990).

The first investigations of L-GLU binding to synaptic membrane fractions derived from the whole rat brain were performed using [^{14}C]-GLU (Michaelis et al. 1974; Roberts 1974). Using [^{14}C]-labeled ligand hampered further progress of receptor binding study owing to the low specific radioactivity of

[^{14}C]-GLU which was about 250–300 mCi/mM. However, these experiments have already shown the Na^+-independent character of L-GLU binding sites, consistent with a finite number of receptor sites. The membranes of the rat cerebral cortex were found to possess a single population of glutamate binding sites which exhibit K_d = 8 µM and B_{max} = 0.03 nM/mg protein, whereas L-GLU binding with the membranes of the whole rat brain was biphasic with a high affinity site (K_d = 0.2 µM and B_{max} = 1.8 nM/mg protein), and a low affinity site (K_d = 4.4 µM and B_{max} = 8.8 nM/mg protein). In proteolipid fractions of the rat cerebral cortex, even three L-GLU binding sites were found: a high affinity site, a medium affinity, and a low affinity ones. Towards the end of the 1970s, when [^{3}H]-labeled ligands of high specific radioactivity (approx. 30 Ci/mM) became available, the study of L-GLU binding sites began to develop more intensively. Using this approach when studying synaptic membranes of the cat cerebellum, two types of binding sites were also revealed (Head et al. 1980). High affinity binding did not depend on the presence of Na^+, and was observed in fractions of plasma and synaptic membranes. L-ASP inhibited L-GLU binding.

The studies of specific binding in the membrane fractions of insect CNS also revealed the presence of receptors sensitive to GLU (Fiedler et al. 1986). Two L-[^{3}H]-GLU binding sites with dissociation constants (K_d) 29 nM and 249 nM, and maximum binding capacities (B_{max}) of 5.7 and 24.6 pM/mg protein were shown. Some amino acid analogs inhibited the process of specific L-[^{3}H]-GLU binding. The existence of two or more types of sites is very interesting, particularly if they could be identified with different types of synaptic receptors.

The greatest number of studies seems to be devoted to L-GLU binding to CNS membranes (Roberts 1981, 1984; McLennan 1981; Foster and Fagg 1984; Dambinova 1989). The binding constants in accordance with their affinity for L-[^{3}H]-GLU obtained from these studies can be divided into three groups: K_d < 100 nM, K_d = 100–1000 nM and K_d > 1000 nM. The majority of investigators admit that high affinity binding sites (K_d < 100 nM) seem to be due to the nonspecific absorption of the label; low affinity binding sites (K_d > 1000 nM) may reflect the processes of L-GLU transport and only the binding sites with K_d = 100–1000 nM are associated with postsynaptic receptors.

Using the radioligand method, it is often difficult to separate the receptor binding sites and the systems of binding to the uptake carrier. Roberts (1974) originally suggested that L-GLU binding studies in the presence of Na^+ might permit visualization of binding to the uptake carrier. It was shown that binding in solution containing Na^+ was not effectively inhibited by some receptor antagonists, while threo-3-hydroaspartate, an efficient uptake inhibitor, was active. Therefore, the study of binding in the presence of Na^+ makes it possible to define the system of high affinity uptake of transmitter amino acids, whereas amino acid binding in the Na^+-free medium may reveal the existence of receptor binding sites on the membranes (Roberts 1974; Dambinova and Gorodinsky 1984; Dambinova 1989).

In conclusion, it should be noted that the presence of receptor binding sites is not by itself direct proof of the existence of a certain type of synaptic neurotransmitter system. Moreover, in spite of the fact that the first studies on receptor binding appeared more than 10 years ago, there is poor correlation between the pharmacological activity of the agonists and antagonists and L-GLU binding sites in many studies. It will be of great interest to see whether the complexity of binding sites is associated with some subpopulations of amino acid receptors and what their functional role is.

3.5 Pharmacology of Amino Acid Receptors

Many amino acids and, primarily L-GLU and L-ASP, have been known, for quite a long time, to exert a depolarizing influence upon the neurons of the central and peripheral nervous systems of vertebrates and invertebrates. However, it has been difficult to obtain compeling evidence to support the putative transmitter role of these substances. One of the main obstacles which made it difficult to solve this problem was the absence of selective blockers which could be used to compare their influence upon the effects of amino acid with effects caused by the action of endogenous neurotransmitter. Only recently, considerable progress in this area has become possible, due to the discovery of selective antagonists that can distinguish between the effects produced by different amino acid agonists (see reviews by Watkins 1978, 1981; Watkins and Evans 1981; Collingridge and Leiter 1983; Foster and Fagg 1984; Fonnum 1984; Roberts 1984; Piotrovsky 1987; Mac Dermott and Dale 1987).

At present, more than 100 agonists of excitatory amino acids are known to exist (Watkins 1981; Piotrovsky 1987). The molecule of almost each of them possesses one cationic and two anionic groups. The anionic groups usually occupy an α-,β-position (e.g., in the ASP molecule), or an α-,γ-position (e.g., in the GLU molecule) in relation to one another (Fig. 54). Kainic acid (KA), quisqualic acid (Q), and N-methyl-D-aspartic acid (NMDA) are known to be the most potent amino acid agonists. Figure 55 shows their structural formulas.

The neuronal structures possess a different sensitivity to amino acid action. Thus, in one of the first studies of this kind, it was found that the sensitivity of different groups of spinal neurons to L-GLU and L-ASP is not identical (Duggan 1974), which could be explained by the presence of separate receptors for these amino acids. In connection with these data, it was suggested that a conformationally restricted glutamate analog, such as KA, might be a more selective agonist for glutamate receptors than L-GLU itself, since the latter can adopt different conformations and thus interact

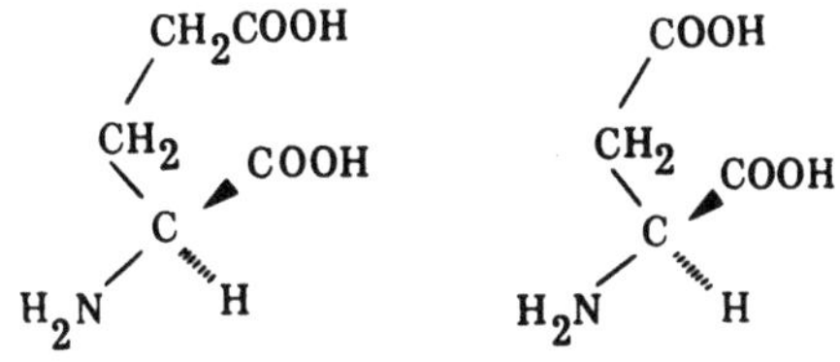

L-GLUTAMIC ACID L-ASPARTIC ACID

Fig. 54. Structures of L-glutamic and L-aspartic acids

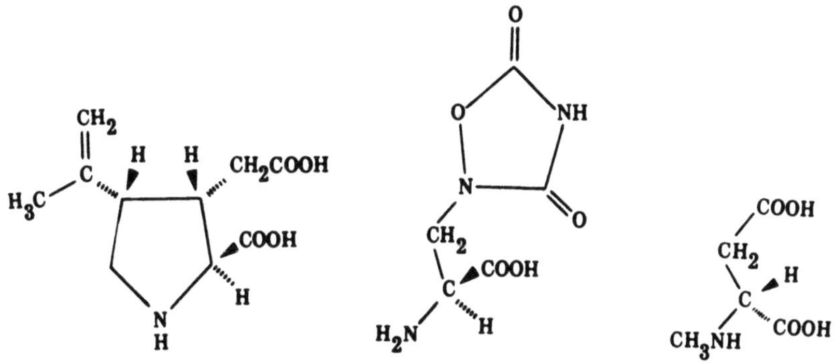

KAINIC ACID QUISQUALIC ACID N-METHYL-D-ASPARTIC ACID

Fig. 55. Structures of amino acids agonists used in studies of receptor types

with more than one type of postsynaptic receptors. At the same time, KA cannot interact with receptors activated by the aspartate analog, NMDA (Johnston et al. 1974). Indeed, the relative sensitivity of different groups of spinal neurons to KA and NMDA varied more markedly than the corresponding potency of L-GLU relative to L-ASP (McCulloch et al. 1974).

The first agents found to differentiate between various amino acid agonists were magnesium ions (Evans et al. 1977; Davis and Watkins 1977; Ault et al. 1980). In the neurons of the frog and cat spinal cord, Mg^{2+} depressed responses evoked by NMDA and had no effect on responses to L-GLU, KA, and Q (Fig. 56). Co^{2+} and Ni^{2+} ions exerted a qualitatively similar but more potent action than Mg^{2+}. The data obtained testified to the existence of at least two types of excitatory amino acid receptors: Mg^{2+}-sensitive and Mg^{2+}-insensitive types. The former were called aspartate-preferring, whereas the latter were referred to as glutamate-preferring. Action similar to that evoked by magnesium ions was produced by a number of monoamino- and diaminodicarbonic acids of chain length longer than that present in the

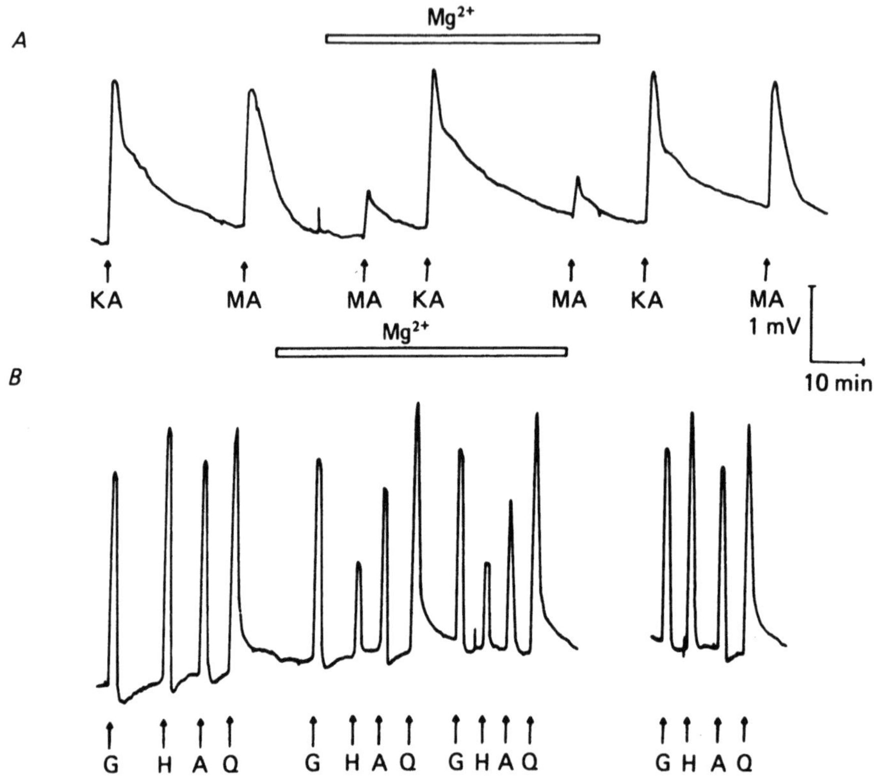

Fig. 56A,B. Selective depression by Mg^{2+} of frog motoneuron depolarizations induced by excitant amino acids. Records show ventral root potentials induced by: **A** *upper series* 12 µM kainate (*KA*) and 40 µM NMDA (*MA*); **B** *lower series* 1 mM L-glutamate (*G*), 100 µM L-homocysteate (*H*), 1.2 mM L-aspartate (*A*), and 5 µM quisqualate (*Q*). Mg^{2+} (1 mM) was present in the medium during the periods represented by the *bars above the records*. The *break in the lower record* represents 20 min. Calibration: *horizontal* 10 min; *vertical* 1 mV. (Ault et al. 1980)

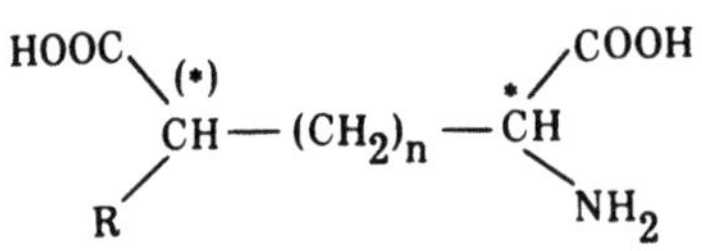

n	R	*	(*)	COMPOUND	ABBREVIATION
2	H	D	-	D-α-AMINOADIPIC ACID	D-α-AA
3	H	D	-	D-α-AMINOPIMELIC ACID	D-α-AP
4	H	D	-	D-α-AMINOSUBERIC ACID	D-α-AS
3	NH_2	D or L		α,ε-DIAMINOPIMELIC ACID	α,ε-DAP

Fig. 57. Structures of long chain monoamino and diamino dicarboxylic acids

glutamate molecule and having the D configuration at the aminated carbon atom(s) (Fig. 57). Responses elicited by NMDA appeared to be most susceptible to these substances, whereas responses induced by KA, L-GLU, and Q were either less susceptible to antagonism or absolutely unaffected (Biscoe et al. 1977a,b; Evans and Watkins 1978; McLennan and Lodge 1979, etc.). These data confirm the existence of two types of excitatory amino acid receptors. The NMDA type of receptors is activated by NMDA and blocked by Mg^{2+}, Co^{2+}, and Ni^{2+}, and also by some organic blockers such as D-α-aminoadipic acid (DAA) and its derivatives. Another type of receptors is activated by L-GLU, KA, and Q, and is resistant to the NMDA antagonists.

The non-NMDA receptors are represented by two subtypes for which KA and Q are the preferential agonists. In cat spinal neurons the iontophoretic application of L-glutamic acid diethyl ester (GDEE) (Fig. 58) blocked Q- (not KA-) responses (McLennan and Lodge 1979; Davies and Watkins 1979). However, the results of the influence of GDEE should be interpreted with some caution, taking into account its weak antagonism. It is supposed that GDEE acts predominantly on responses induced by amino acids that are subject to rapid uptake. Therefore its action may be confined to near the tip of the administering micropipet (Davies and Watkins 1979; Watkins 1981). The uptake of other amino acids, such as NMDA and KA, is slower (Balcar and Johnston 1972; Johnston et al. 1979). Therefore, they may influence receptors which are outside the range of GDEE action. Moreover, GDEE has little, if any, influence on amino acid-induced responses during bath perfusion of the substance in the isolated spinal cord (Evans et al. 1979).

Besides the differentiated effect of GDEE on KA- and Q-induced responses, numerous other data support the hypothesis of the existence of KA and Q receptors. KA and Q relative potencies were shown to be different in different preparations. For example, in the frog and rat spinal cord, Q is more effective than KA (Biscoe et al. 1976), whereas on isolated dorsal roots of the immature rat, the relation is reverse (Davies et al. 1979). KA-induced responses were much more suppressed by 2-amino-4-phosphonobutyrate (APB) than Q-induced responses (Davies et al. 1979).

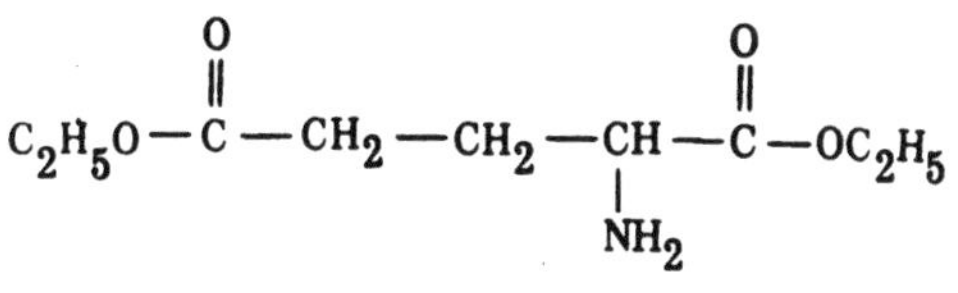

Fig. 58. Structure of L-glutamic acid diethylester

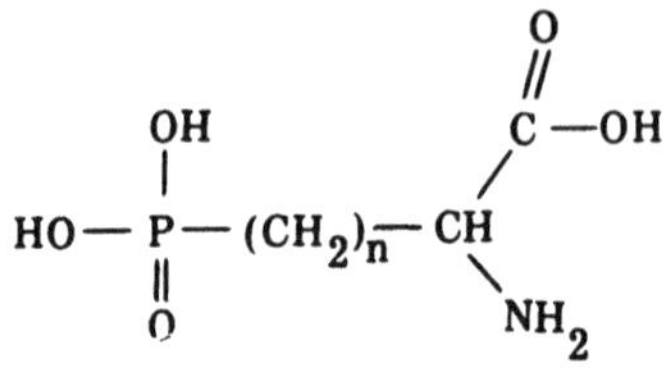

n	COMPOUND	ABBREVIATION
1	2-AMINO-3-PHOSPHONOPROPIONIC ACID	2 APP
2	2-AMINO-4-PHOSPHONOBUTIRIC ACID	2 APB
3	2-AMINO-5-PHOSPHONOVALERIC ACID	2 APV

Fig. 59. Structure of phosphono amino carboxylic acids

The ω-phosphonic analogs of glutamic and aspartic acids (Fig. 59) antagonized the responses evoked by amino acid excitants. For example, APB in both frog and cat spinal neurons similarly inhibited KA- and NMDA-induced responses and affected Q-induced responses less (Evans et al. 1979; Davies and Watkins 1979). It should be noted that APB depressed only responses elicited by amino acid agonists and did not affect cholinergic synapses, in contrast to GDEE antagonism.

The [±]-APB blocking activity is carried out by the [−]-isomer, whereas the [+]-isomer is excitatory (Watkins 1981). It is curious that a phosphonate with a longer chain than that of an APB molecule, 2-amino-5-phosphonovalerate (APV), i.e., the ω-phosphono analog of D, L-α-aminoadipate, seems to be the most potent NMDA antagonist found so far. Even its racemic mixture is 15 times stronger than DAA, and is over 100 times more potent than APB. Another important difference in APV influence as compared to APB is a considerably greater selectivity.

The organic antagonists of close structural similarity to excitatory amino acid agonists suggested a competitive mechanism in their action, which has been supported by binding data (Watkins and Olverman 1987). Recently, however, it has been shown that the effect of a number of substances with structures quite different from those of competitive NMDA antagonists can be explained in terms of reversible blocking of open channels of synaptic receptors. These noncompetitive antagonists include phencyclidine (PCP) ketamine, MK-801, dexoxadrol, MLV 5860, LY 154045, and some others. Such components are thought to act at the level of the NMDA receptor-associated ion channel. In fact, the block by these compounds of excitatory amino acid responses was highly voltage-dependent (Fig. 60), the presence of the agonist enhanced the blockade, and these antagonists did not influence the binding of the competitive antagonist APV or the binding of NMDA agonists (Pennfather and Quastel 1982; Foster and Fagg 1984; Harrison and Simmonds 1985; Martin and Lodge 1985; MacDonald et al. 1987; Kemp et

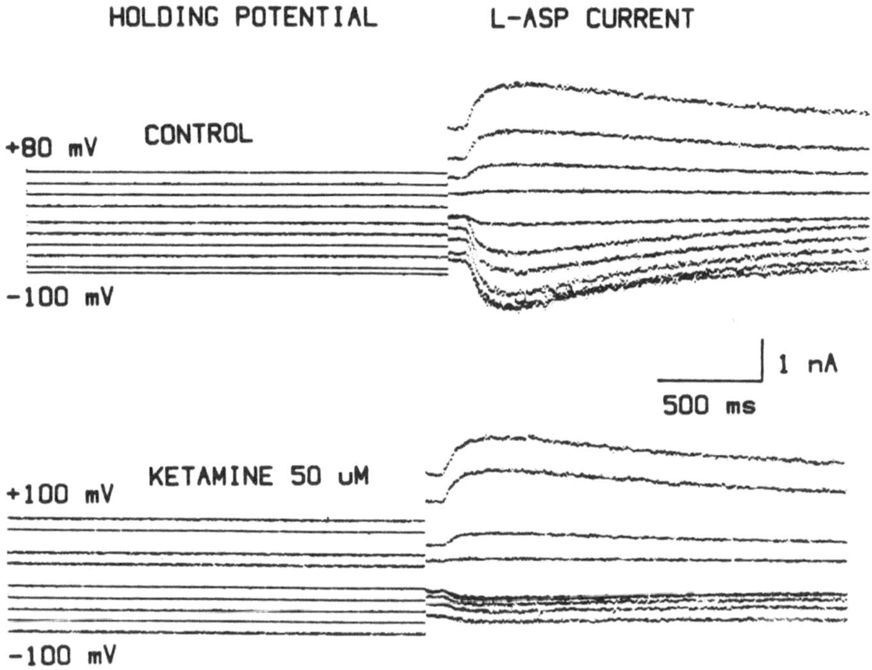

Fig. 60. The voltage-dependent block of inward L-aspartic acid (*L-Asp*) currents in cultured neurones by ketamine (50 M). On the *left* are superimposed voltages, and on the *right* L-Asp currents. In the *upper panel* a single application of L-Asp (50 M) was made at a series of ten holding potentials with a sufficient interval between each application to minimize desensitization. Following the addition of Ketamine to the bath (*lower*), the inward currents evoked by L-Asp were almost eliminated, whereas outward currents were markedly less depressed. L-Asp was repeatedly applied at −70 mV following addition of ketamine to develop a steady-state block. (MacDonald et al. 1987)

al. 1987; Lodge and Johnson 1990). Hence these compounds do not act like classical allosteric modulators of the neurotransmitter binding site, and the block is not affected by the simple occupation of the recognition site, but rather is dependent on the activation of the NMDA channel.

Although the existence of at least three types of amino acid postsynaptic receptors, Q, KA, and NMDA receptors, was generally supported by electrophysiological and binding studies (Foster and Fagg 1984; Monaghan et al. 1989; Headly and Grillner 1990), some refinements of this receptor scheme were needed in the light of currently available data. As a result, some authors suggested a fourth type of receptor, L-APB, or L-AP4-sensitive receptors (Foster and Fagg 1984), since in a number of preparations NMDA antagonists were ineffective with respect to L-APB-sensitive synapses, and responses evoked by KA and Q were insensitive to doses of L-APB, which give a maximal depression of synaptic responses (Evans et al. 1982; Ganong and Cotman 1982; Collins et al. 1983). Moreover, binding studies clearly indicate that Q is not a specific agonist and interact with KA, NMDA, and L-GLU binding sites. At the same time, a structural analog of Q, D, L-α-amino-3-hydroxy-5-methyl-4-isoxazolepropionic acid (AMPA) is more specific for its receptor and GDEE-sensitive. Therefore this receptor subtype was renamed Q/AMPA, or simply AMPA receptor to designate the receptor at which Q potently depolaries neurons (Monaghan et al. 1989; Watkins et al. 1990; Young and Fagg 1990). The fifth subtype of excitatory amino acid receptor, referred to as a metabotrophic, or ACPD receptor, can be activated by Q, isobotenate, and *trans*-1-amino-cyclopentane-1,3-dicarboxylic acid (ACPD) (Monaghan et al. 1989; Watkins et al. 1990; Young and Fagg 1990). Evidence concerning the classification of excitatory amino acid receptors is given in Table 1.

The scantiness of the amino acid receptor pharmacology and the lack of specific ligands for different types of postsynaptic receptors make the use

Table 1. Excitatory amino acid receptors. (Watson and Abbott 1990)

Nomenclature	NMDA	AMPA	Kainate	AP-4	Metabotropic
Previous names	–	Quisqulate	–	L-APB	–
Selective agonists	NMDA	AMPA	Kainate[a]	L-AP4	Quisqualate[b] Ibotenate[b]
Selective antagonists	D-AP-5 CPP CGS19755 CGP37849 Noncompetitive (Glycine site)[c] HA966 7-chlorokynurenate	CNQX DNQX NBQX GAMS	CNQX DNQX NBQX GAMS	–	–
Channel blockers	MK-801 Phencyclidine Ketamine SKF10047	–	–	–	Not applicable
Radioligands	[^{3}H]D-AP5 [^{3}H]CPP [^{3}H]CGS1975 [^{3}H]GLU	[^{3}H]AMPA [^{3}H]CNQX [^{3}H]GLU	[^{3}H]kainate	–	–

[a] Selective to C-fiber only.
[b] Selective in terms of IP_3/DG stimulation only.
[c] Glycine binding site needs to be ocupied for receptor activation.
AMPA, D,L-α-amino-3-hydroxy-5-methyl-4-isoxazole propionic acid; L-AP4 (L-APB), 2-amino-4-phosphonobutanoate; D-AP5, D-amino-5-phosphonopentanoate; CGS19755, 4-phosphonomethyl-2-piperidinecarboxylic acid; CGP37849, D,L-(E)-2-amino-4-methyl-phosphono-3-penthanoic acid; CNQX, 6-cyano-7-nitroquinoxaline-2,3-dione; CPP, (±)-2-carboxypiperazine-4-yl-propyl-1-phosphonic acid; DNQX, 6,7-dinitroquinoxaline-2,3-dione; GAMS, γ-aminomethysulphonate; HA966, 3-amino-1-hydroxypyrrolid-2-one; MK801, (+)-5-methyl-10,11-dihydro-5H-dibenzo(a,d)cyclohepten-5,10-imine maleate; NBQX, 6-nitro-7-sulfamobenzo(f)quinoxaline-2,3-dione; SKF10047, (±)-N-allylnormetazocine.

of the natural ligands as receptor probe very promising. Unfortunately, most of them have an affinity for a number of binding sites. However, some progress has been already made in this direction (Jackson and Usherwood, 1988; Jackson and Parks 1989). This concerns detection, isolation, and study of the properties of neurotoxins in different species of spiders (Kawai et al. 1982; Magazanik et al. 1986). In particular, a toxin (argiopine) isolated from *Argiope* spider venom was shown to suppress the excitatory postsynaptic potential (EPSP) and postjunctional L-GLU-potential at neuromuscular junctions on insect skeletal muscle (Fig. 61). The influence of argiopine was relatively specific and was carried out by two molecular mechanisms: an open channel block (the main mechanism of antagonism by argiopine) and a closed channel block. The interaction of this toxin with the glutamate binding sites on the receptor is considered to be unlikely. Unfortunately, argiopine can also interact with acetylcholine receptors. It can be expected that clarification of the mode of action of neurotoxin and artificial synthesis of its analogs will serve as a reliable tool for analyses of postsynaptic amino acid receptors.

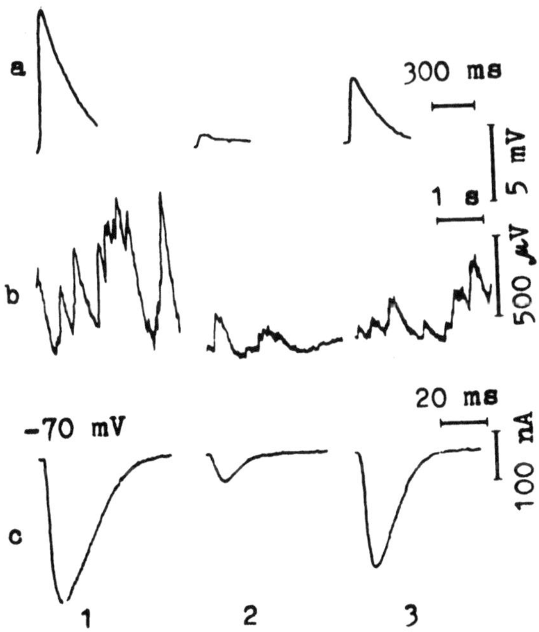

Fig. 61a–c. Effects of toxin of spider *Argiope lobata* (5 μg/ml) on excitatory postsynaptic potential (**a**), miniature excitatory postsynaptic potential (**b**), and excitatory postsynaptic current (**c**) in muscle fibers of fly. *1* control; *2* after application of toxin (80 min); *3* recovery (60 min). (Magazanick et al. 1986)

3.6 The Mode of Action of Excitatory Amino Acids

Information about the responses of neuronal membranes to excitatory amino acids has been gained in many ways: through investigations of membrane potentials and conductance changes, ion fluxes, and metabolic alterations such as increases in cyclic nucleotide levels. L-GLU and its agonists have been known for more than 30 years to act at the postsynaptic level and exert a powerful depolarizing effect. At the same time, no great conductance changes in cell membrane was noticed as a rule. In cat rubrospinal neurons (Altman et al. 1976), or lamprey Müller cells (Wickelgren 1977), increases in membrane conductance by iontophoretic L-GLU application were considerably smaller than those by corresponding doses of GABA. No changes in membrane conductance were produced by iontophoretic application of L-GLU either in frog spinal motoneurons (Shapovalov et al. 1978) or in cat spinal neurons (Lambert et al. 1978). Depolarizations with conductance decreases have been observed in tissue culture by microperfusion of L-GLU and its analogs in the mouse spinal neurons (MacDonald et al. 1982). The conductance increases evoked by excitatory amino acids were attributed by some authors to the uptake of these substances (Cox et al. 1977; Engberg et al. 1979), although this explanation is hardly applicable to KA, which is not taken up by presynaptic endings (Johnston et al. 1979). Thus the effects of excitatory amino acids are conditioned not so much by the mechanism of uptake as for other reasons. The explanation for small conductance increases and large depolarizations by excitatory amino acids may be large driving forces, the difference between the membrane potential, and the reversal potential for EPSPs and for L-GLU-induced depolarization. All these factors could compensate for the small conductance increases by glutamate (Martin et al. 1970; Martin and Ringham 1975; Altman et al. 1976; Wickelgren 1977; Homma 1983). In contrast, the driving force for inhibitory postsynaptic potentials (IPSPs) and glycine or GABA is small, and the conductance increases should be large for effective response.

The ionic mechanisms of membrane conductance changes are also different in different objects. For example, in Müller cells of lamprey (Matthews and Wickelgren 1979), the ions contributing to an increase in membrane conductance by L-GLU and EPSPs were Na^+ and K^+. Reversal potentials for responses evoked by nerve stimulation and L-GLU application were identical and were averaged of the equilibrium potentials of Na^+ and K^+. The role of Ca^{2+} in the responses for L-GLU and EPSPs in this object is very small, as in the neuromuscular synapse of crayfish (Onodera and Takeuchi 1975).

Sodium dependence of L-GLU responses was demonstrated for frog motoneurons (Shapovalov et al. 1978), mouse spinal neurons in tissue culture (Ransom et al. 1977), pyramidal cells of the hippocampus (Hablitz and Langmoen 1982), and for many other objects. When Na^+ concentration was reduced, there were negative shifts in reversal potentials of glutamate response. Increases in K^+ in perfusing fluid resulted in positive shifts in reversal potentials for L-GLU responses in frog motoneurons (Shapovalov et al. 1978). The data obtained explained the observed conductance changes evoked by L-GLU application by Na^+ conductance increases and K^+ conductance reductions (Shapovalov et al. 1978; Engberg et al. 1979). It should be pointed out, however, that the presumed K^+-conductance reduction under the action of amino acids correlates badly with the observed elevation of extracellular K^+ concentration. Using ion-selective microelectrodes made it possible to detect increases in outer K^+ concentration, suggesting potassium ion efflux from the cells during L-GLU responses (Sonnhof et al. 1978; Sonnhof and Buhrle 1980). Therefore amino acid effects on postsynaptic receptors should be distinguished from depolarization evoked by an elevated concentration of K^+ released from the cells.

Probably the general property of glutamatergic neurons lies in discrepancies between reversal potentials of EPSPs and those of amino acid potentials. Such data exist for giant spinal motoneurons of the lamprey (Homma 1983), frog motoneurons (Shapovalov et al. 1978), mouse spinal neurons in tissue culture (Ransom et al. 1977), spinal neurons in vivo (Curtis and Watkins 1965), etc. It is therefore suggested that the mechanism of excitatory amino acid action is more complicated than a simple interaction with membrane receptor and a change in membrane conductance evoked by this process. In particular, the existence of differences in conductance-producing mechanisms or receptors for EPSPs and excitatory amino acids, as well as differences in the electrotonic length between the synaptic locations activated by nerve stimulation and the location of the amino acid electrode for iontophoresis is assumed (Ransom et al. 1977; Matthews and Wickelgren 1979).

As stated above, several types of excitatory amino acid receptors are recognized by selective agonists and antagonists. The introduction of methods applying intracellular perfusion, concentration clamp, noise analysis, and single channel recording in primary cultures was an important step in the pharmacological dissection of amino acid-induced responses and has radically changed the situation in understanding the basic mechanisms of synaptic transmission. Of special interest is the voltage-clamp analysis of membrane currents elicited by the application of different analogs of excitatory amino acids. As shown in the studies on the neurons of rat hippocampus performed by intracellular perfusion technique in combination with the method of concentration clamp (Kiskin et al. 1986a,b; Kiskin and Tsyndrenko 1985), by the action of L-GLU-, KA-, and Q-containing solutions (0.05–25 mM), transient inward ionic currents were observed (Fig. 62). These currents, carried mainly by Na^+ and K^+, were linearly dependent upon the potential. The reversal potential for conductances linked to L-GLU, KA, and Q were similar, making it possible to suggest that the agonists tested activate the same excitatory membrane receptors. It is curious that application of L-GLU and Q produced a rapid and complete desensitization of ionic currents, whereas application of KA resulted in the initiation of inward current but

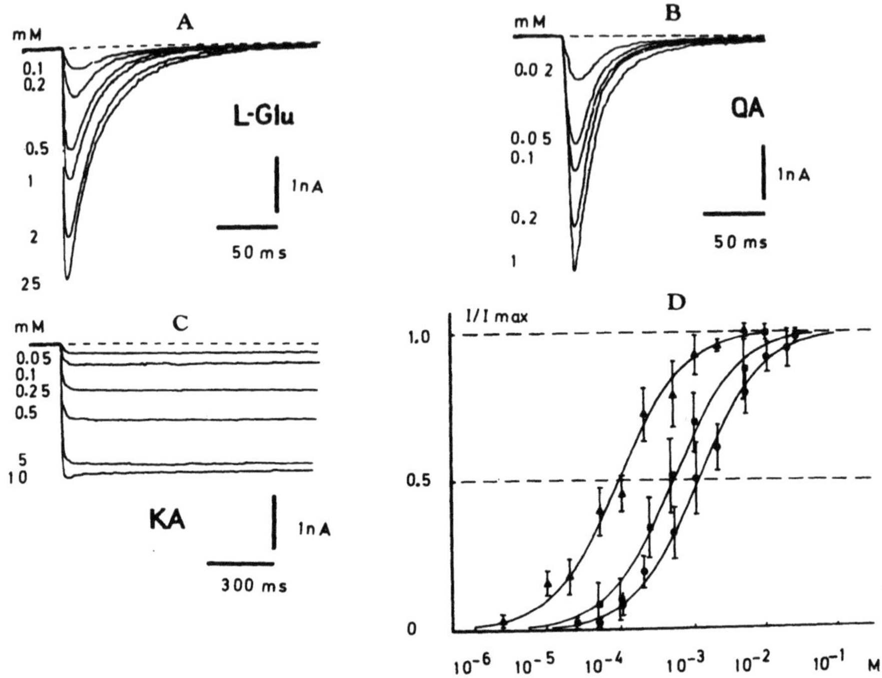

Fig. 62A–D. Inward currents elicited by L-Glu (**A**), quisqualate (*QA*) (**B**) and kainate (*KA*) (**C**) in rat hippocampal neurons. Solutions containing given concentrations of agonists (indicated near corresponding current curves) were applied in a step-wise manner and maintained around the cell during the measurement period. The moment at which the aplication started is clearly visible from the beginning of inward current activation. Recordings **A–C** were obtained from different neurons. Holding voltage = 100 mV. **D** Dose-response relationships for QA (four cells), KA (five cells), and L-Glu (14 cells). Peak values of currents elicited by application of each substance at various concentrations were measured and normalized to the maximum current. *Smooth curves* are single-site binding isotherms with K_d values of 9.3 10^{-5} M (QA), 5.0 10^{-4} M (KA), and 1.1 10^{-3} M (L-Glu). These K_d values were calculated to obtain the best fit using the least-squares method. *Vertical bars* at each point indicate ± SEM. (Kiskin et al. 1986b)

responses were not desensitized even by a prolonged action of KA. It cannot be ruled out that such peculiarities of reaction on the application of KA, together with a low effectiveness of uptake systems for KA explain a strong long-lasting depolarizing and neurotoxic influence of this substance on CNS structures (Coyle 1983). Experiments in which Q was applied after a complete desensitization evoked by L-GLU and vice versa showed complete and reversible desensitization. In a similar case with KA, a normal response to L-GLU could be caused after exposure of KA, providing the lack of desensitization produced by the latter substance. This suggests that all the agonists are targeted on the same membrane receptor. It should also be noted that the above study, as opposed to the conventional opinion that L-GLU and L-ASP are mixed agonists of excitatory amino acids, showed the

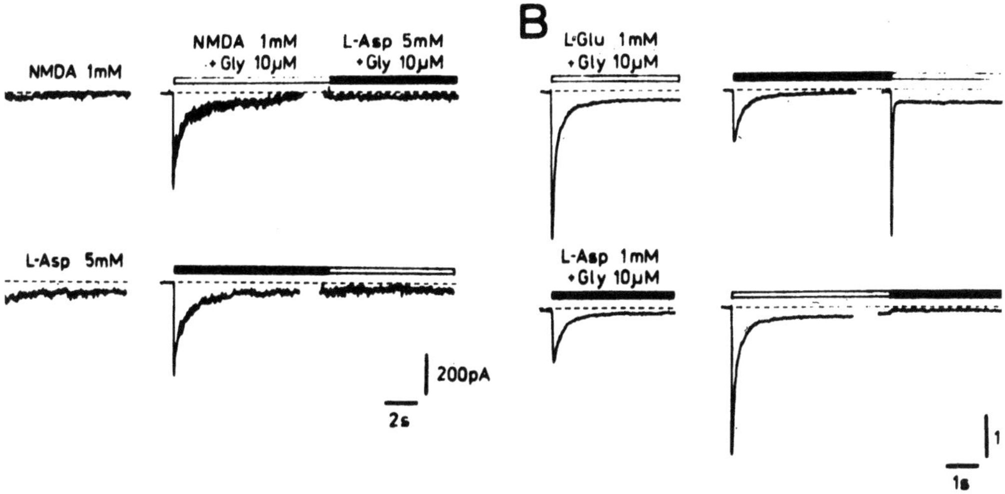

Fig. 63A,B. The responses of hippocampal neurons to NMDA and other agonists, and their cross-desensitization. Application programs are represented by *horizontal bars*. **A** Ionic currents elicited by NMDA and L-Asp (*left panel*) were enhanced when agonists were applied with 10 μM glycine (*Gly*) to the same neuron (*right panel*). After incubation in the agonist-containing solution for 30 s (the *gaps* in current traces), no further response to the other agonist could be obtained. **B** L-Glu response elicited on the background of 10 μM Gly had a slow (NMDA) component (*top left*). This component was completely depressed after application of L-Asp and 10 μM Gly, while the rapid (non-NMDA) component persisted (*top right*). The response to L-Asp and Gly in the same neuron (*bottom left*) was completely desensitized by application of L-Glu with 10 μM Gly (*bottom right*). The peak of L-Glu-induced current was truncated in all records because its amplitude was more than five times that of NMDA component. **A** and **B** Different neurons; holding potential = 100 mV. (Chizhmakov et al. 1989)

existence of receptors insensitive to L-ASP and NMDA activated by L-GLU, Q, and KA. However, in recent electrophysiological experiments, the same research group (Chizmakov et al. 1989) was able to obtain NMDA- and L-ASP-activated ionic currents using the effect of glycine potentiation in Mg^{2+}-free extracellular solutions (Fig. 63).

The results obtained are interesting, but it does not seem possible to extend the conclusions drawn from the above study to other objects. This is supported by a recent study on embryonic chick motoneurons maintained in cell culture in combination with voltage-clamp technique (O'Brien and Fischbach 1986). It was shown that during voltage clamp on the membrane at −50 mV, L-GLU ejected through the micropipet evoked an inward current. However, this current induced by $10^{-3} - 10^{-4}$ M L-GLU was not desensitized as in Kiskin's work (Kiskin et al. 1986a,b), but rose rapidly during a 400-ms pressure pulse, and then decayed exponentially. During a prolonged application of L-GLU (20 s) the ionic current gradually decayed; the mean decrease in current was only 17.4% of the peak level, testifying to an extremely weak desensitization. Dose-response curves were constructed with

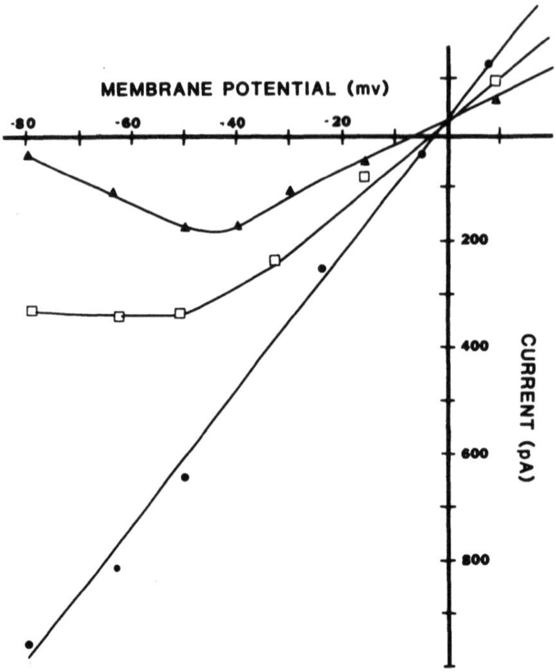

Fig. 64. Voltage-current relations in the presence of excitatory amino acid agonists in embryonic chick motoneuron. Same motoneuron was tested with NMDA (*closed triangles*), glutamate (*open squares*), and kainate (*closed circles*). (O'Brien and Fischbach 1986)

L-GLU, L-ASP, NMDA, KA, and Q; in each case the curve followed a simple hyperbolic curve without any indication of the multiple binding sites or cooperativity (Fig. 64). The reversal potentials of the various agonists were practically identical (about −6 mV). Although dose-response curves for L-GLU and its agonists and the identical reversal potentials of the substances investigated were consistent with a single site of action, experiments with membrane hyperpolarization suggested the existence of more than one receptor-ionophore complex. The responses activated by L-GLU did not increase with hyperpolarization. Under the same conditions, KA-induced current increased with hyperpolarization, whereas the NMDA current decayed. The data obtained seem to testify that L-GLU activated both types of channel. Indeed, the first type of channel was activated by NMDA and L-ASP and inhibited by APV. The second type of channel was activated by KA and Q and was not inhibited by APV. When membrane potential was −50 mV, 38% of L-GLU current was due to the activation of the first type of receptor and 62% to the second type. No proof was obtained that the second type of receptor can discriminate between KA and Q, suggesting that in motoneurons KA and Q may interact with the same receptor type. In the above study, γ-glutamylglycine equally well blocked responses evoked by KA and Q. On the contrary, GDEE had no effect on Q, KA, or L-GLU currents. The above data show the existence of different mechanisms of excitatory amino acid in different preparations.

All classes of excitatory amino acid receptors differ in their electrophysiological and pharmacological properties as well as in details of their anatomical organization. Excitatory amino acids can produce both a rapid and a more persistent excitatory effect on the same postsynaptic membrane. Thus EPSPs mediated by transmitters acting at non-NMDA receptors exhibit fast rise times (1–3 ms) and have decay times of around 10–50 ms. Synaptic currents underlying fast EPSPs have rise and decay times of less than 1 ms (Nelson et al. 1986; MacDermott and Dale 1987). EPSPs mediated by a transmitter acting at NMDA receptors are of longer duration, with a rise of 20–25 ms and a total duration of 200–300 ms (Jahr and Yoshioka 1986; Thomson 1986; MacDermott and Dale 1987). Accordingly, EPSPs recorded from the diverse regions in the CNS have a dual nature and consist of a summation of fast non-NMDA and slow NMDA receptor-mediated components.

NMDA and non-NMDA receptors can also be distinguished by their sensitivity to Mg^{2+}. Addition of Mg^{2+} can effectively reduce the NMDA-induced responses; and a total block is obtained with 1–2 mM concentrations. The non-NMDA responses are not influenced by the blocking action of Mg^{2+} (Ault et al. 1980; Ascher and Nowak 1987). Mg^{2+} evidently acts by blocking the NMDA ion channel in a voltage-dependent manner (Nowak et al. 1984; Mayer et al. 1984; Ascher and Nowak 1987). The blocking action of Mg^{2+} is more pronounced at the resting membrane potentials, but progressive depolarization causes a gradual aleviation of the channel blockade, creating a region of negative slope conductance in the current-voltage relation of the neuron, and can lead to a regenerative depolarization. KA and Q activate relatively voltage-insensitive conductances, whereas L-GLU and several of its analogs act as mixed agonists that activate conductances associated with both NMDA and non-NMDA receptors.

NMDA and non-NMDA receptors also differ in the ion permeability of their associated channels. Both of them are permeable for K^+ and Na^+; however, the reversal potentials of non-NMDA-induced responses are more sensitive to changes in extracellular Na^+. Moreover, changing the concentration of external Ca^{2+} strongly alters the reversal potential of NMDA but not non-NMDA-induced responses (Ault et al. 1980; Ascher and Nowak 1986, 1987). These findings suggest that the NMDA channel may be permeable to Ca^{2+}. The non-NMDA channel, in contrast, seems to exhibit low permeability to Ca^{2+} (McDermott and Dale 1987).

Three types of excitatory amino acid receptors have usually been associated with three types of cation-selective channels, which differed in their kinetic and amplitude characteristics. However, recent studies at the single channel level suggest the existence of another mechanism (Jahr and Stevens 1987; Cull-Candy and Usowicz 1987). According to the above authors, the glutamate receptor is a single receptor-channel complex which may be coupled to both NMDA and non-NMDA receptors. In this case, the effects of NMDA, KA, and Q are mediated by a family of channel openings with a range of conductance changes, ionic permeabilities, and activation kinetics. Each agonist activates both the main conductance and rare events of a smaller or a larger size. Therefore, the selectivity of each agonist is defined by a unique spectrum of activation of various conductance states. This hypothesis contradicts the observation showing that NMDA, KA, and Q binding exhibits different regional localizations and effects of non-competitive antagonists like Mg^{2+}, ketamine, phencyclidine, and MK-801, which block channels associated with NMDA receptors without any effect on kainate receptor channels. A more conservative hypothesis suggests the existence of three distinct and independent receptors, but assumes that each type of membrane receptor is coupled with a similar channel that can adopt a variety of conductance states.

Moreover, excitatory amino acids have been shown to increase cAMP and cGMP levels in intact preparations of nerve and muscle tissue (Ferrendelli et al. 1974; Kinscherf et al. 1976; Garthwaite and Balazs 1979). The effects observed are Ca^{2+}-dependent and are also enhanced in the presence of depolarizing agents. The stimulation of cyclic nucleotide synthesis is assumed to be a secondary process following depolarization and associated with Ca^{2+} influx. This process is determined by the morphological integrity of tissue, suggesting a possible involvement of cyclic nucleotides as an intracellular second messenger for any postsynaptic action of excitatory amino acids.

The data presented in this chapter suggest that the mechanisms of amino acid-using synapses differ considerably from the mechanisms of "classic" neuromuscular synapses. This refers to the mechanisms of transmitter release, its interaction with synaptic receptors, cessation of transmitter action, etc. At the same time, it is well established that L-GLU plays the central metabolic role in the brain; it is involved in synaptic transmission in the central and peripheral nervous system in the processes of learning and memory, and it is beneficial in several neurological disorders. Some other functions have also been ascribed to L-GLU. All this makes the study of the powerful and excitatory action of L-GLU in synaptic transmission very important from both theoretical and practical points of view.

Chapter 4

Pharmacology of Synaptic Transmission in Receptors of the Acousticolateral System

The sensory epithelia of the inner ear of mammals and lateral-line system of fish and aquatic amphibia contain mechano- and electrosensitive hair cells. These cells release a chemical transmitter both at rest and during activation. In turn, the chemical transmitter activates specialized postsynaptic receptors, which leads to the depolarization of afferent nerve fibers. On the one hand, one chemical substance is believed to carry out the neurotransmitter function. On the other hand, during the stimulation of sensory organs other chemicals are also probably released from hair and supporting cells – neuromodulators. These chemicals may act on receptors located on the hair cells or afferent fibers to alter responsiveness to the neurotransmitter. For a long time, the possible role of putative neurotransmitters such as acetylcholine (ACh), 5-hydroxytryptamine, norephinephrine, dopamine, serotonine, glycine, taurine, and gamma-aminobutyric acid (GABA) was elucidated. Lately, great interest has been focused on the idea that excitatory amino acids (L-GLU and L-ASP) or a structurally related compound possibly play a role in synaptic transmission in the receptors of the acousticolateral system. These substances are known to act as neurotransmitters in a number of synapses in the peripheral and central nervous systems of invertebrates and vertebrates (Gerschenfeld 1973; Krnjevic 1974). Many reviews on this topic have been published (Fex 1973; Guth et al. 1976, 1981, 1985; Klinke 1981, 1986; Guth and Melamed 1982; Bobbin et al. 1984, 1985; Bledsoe 1986; Bledsoe et al. 1988; Akoev and Andrianov 1989).

4.1 The Organ of Hearing

The role of EAA in synaptic transmission was studied in great detail on hair cells of the organ of Corti. In electrophysiological experiments, the perfusion of scala tympani with L-GLU and L-ASP at a concentration of 10^{-2}–10^{-3} M has been shown to substantially increase the firing rate of the afferent fibers and to reduce sound-evoked responses. This was accompanied by a decrease in the amplitude of the compound action potential (CAP) of the auditory nerve without change in the cochlear microphonic (CM) (Fig.

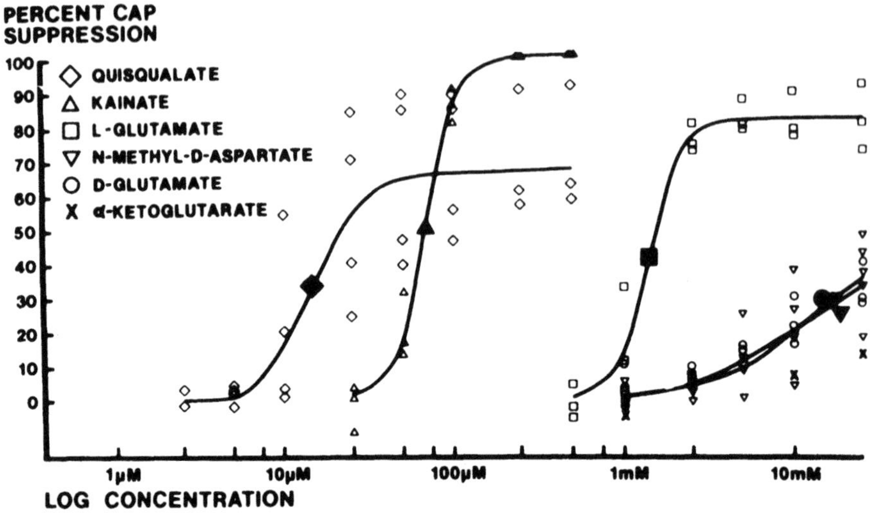

Fig. 65. Dose-response functions illustrating the percentage suppression of mean compound auditory nerve action potentials (*CAP*) amplitude produced in guinea-pig cochlea by perfusing each concentration of agonist. *Open symbols* (agonist designations in *upper left*) indicate individual experimental values. *Solid lines with solid symbols* indicate agonist dose-mean response functions derived from the best fit of associated individual values. All data represent averages of the suppression observed across all effective stimulus intensities rather than effects produced on responses evoked by any given stimulus intensity. Note that fitted functions for quisqualate, kainate, and L-glutamate approach asymptote within the range of test concentrations (2.5 M–25 mM), whereas those for D-glutamate, N-methyl-D-aspartate, or ketoglutarate (control) do not. (Reprinted with permission from Comp. Biochem. Physiol. v. 84C, Jenison et al. Comparative actions of quisqualate, excitatory amino acid agonists, on guinea-pig cochlear potentials. Copyright 1986, Pergamon Press PLC.)

65) (Klnke and Oertel 1977a; Bobbin and Thompson 1978; Bobbin 1979; Comis and Leng 1979; Fex and Martin 1980; Bobbin et al. 1985; Jenison et al. 1986). A rise in the resting firing rate in the fibers of the auditory nerve and a decrease in the amplitude of the CAP allowed the suggestion that in the cochlea there are membrane receptors sensitive to L-GLU and L-ASP. The reduction of CAP induced by L-GLU and L-ASP was due to an increase in discharge rate of the afferent nerve fibers, making them less responsive to sound.

A detailed study was carried out to elucitade the possible role of L-GLU in synaptic transmission in hair cells in the guinea pig cochlea (Gleich et al. 1990) L-GLU injection (2 mM) in to scala tympani produced a variety of effects: decrease of the toneevoked activity without a change in the resting firing rate; reduction in resting activity without a preceding increase, increase in resting activity, reduction of evoked discharge rate belov spontaneous activity. Thus, the effects of external application of L-GLU are complex which lead authors to the conclusion that L-GLU is not the excitatory afferent neurotransmitter in the inner hair cells, although this amino acid may have a role a modulator or cotransmitter. The putative neurotransmitter role of L-GLU was confirmed further by experiments with the effect of degradative enzyme for glutamate, glutamate dehydrogenase, on guinea-pig cochlea (Rebillard and Bryant 1989). The glutamate dehydrogenase was shown to decrease in the auditory nerve responsiveness in a dose-dependent manner. This result indicates that the neurotransmitter pool of L-GLU is present in the synapse of guinea-pig cochlea.

KA, a conformationally restricted analog of L-GLU, was established as exerting a potent depolarizing effect on various neurons. KA was found to produce a dose-dependent reduction in the amplitude of CAP in the guinea-pig cochlea without affecting the CM and the summating potential (SP) (Fig. 65) (Bledsoe et al. 1981a; Kusakari et al. 1984; Jenison et al. 1986). These data showed that KA excited the afferent fibers due to the activation of the postsynaptic receptors. This has been confirmed by Pujol et al. (1985), who showed that KA produced a selective swelling of afferent nerve terminals (auditory dendrites) contacting with inner cells in adult guinea-pig and rat cochlea.

The presence of the kainate binding protein was shown in the frog inner ear by means of monoclonal and polyclonal antibodies (Dechesne et al. 1991).

Another agonist of excitatory amino acid receptors, Q, reduced the CAP amplitude without affecting the CM in the guinea-pig cochlea (Fig. 65) (Jenison and Bobbin 1986; Jenison et al. 1986). Moreover, Q increased the firing rate of single afferent nerve fibers. The effectiveness of Q is about two orders of magnitude more potent than L-GLU and four times as potent as KA. It is worth noting that AMPA a potent agonist at the formely known Q ionotropic receptor completely abolished the CAP of the guinea pig cochlea (Puel et al. 1991). In contrast, NMDA exerted no effect either on the CAP or on the CM (Jenison et al. 1986) (Fig. 65). From these findings it appears that postsynaptic receptors for Q and KA but not for NMDA are present in the afferent fibers in mammalian cochlea.

This conclusion is consistent with the results of experiments on the effects produced by the antagonists of amino acid receptors. Venom from the spider *Argiope trifasciata*, a highly specific antagonist of glutamate receptors, suppressed both resting and evoked activity of the primary auditory afferent neurons of the guinea-pig cochlea (Cousillas et al. 1988). The authors believed that the spider venom acted as an open channel blocker.

DLAA was shown to produce no effect on the CAP and CM (Fex and Martin 1980; Bobbin et al. 1981a). The more potent NMDA-preferring antagonists, such as APV, APB, and DL-aminosuberic acid, produced little if any change in guinea-pig cochlear potentials (Bobbin et al. 1981a; Puel et al. 1989). In contrast, the Q-preferring antagonists, such as GDEE, reduced CAP and CM by 86 and 60%, respectively. The broad-spectrum excitatory amino acid antagonist PDA produced a reduction in CAP by 33% and had no effect on the CM, whereas DGG was less effective. Bobbin and Ceasar (1987) have recently examined the effect of another two broad-spectrum antagonist, KENYA, and γ-D-glutamyl-aminomethylsulfonic acid (GAMS). KENYA was shown to be the most potent amino acid antagonist tested to date to suppress the CAP about 50% at 2.5 mM.

It is worth noting that new excitatory amino acid antagonists have recently become available. One of these compounds, 1-(p-bromobenzoyl)-piperazine-2, 3-dicarboxylic acid (pBB-PzDA), has been shown to be more potent and selective than KENYA for non-NMDA receptors (Puel et al. 1989). This antagonist reversibly suppressed the CAP amplitude and increased the latency of the first negative wave of CAP while having no detectable effect on presynaptic potential such as CM and SP. Moreover, the pBB-PzDA reversibly suppressed the firing of single auditory nerve ganglion cells. On the contrary, APB had no effect on the CAP magnitude of the cochlear nerve, N_1 latency, CM, or SP of guinea-pig cochlea (Puel et al. 1988). APB receptors appear not to be involved in neurotransmission between hair cells and afferent nerve fibers.

Thus, available pharmacological data are consistent with the view that in the cochlea NMDA receptor appears to be absent (Bobbin et al. 1984; Jenison et al. 1986; Klinke 1986; Puel et al. 1989), and that an endogenous neurotransmitter acts on KA or Q receptors.

However, recent publication of Puel et al. (1991), Ehrenberger and Felix (1990) provide evidence for the presence of NMDA receptors in the guinea pig cochlea since APV in concentration of 5 mM induced a reduction of the CAP by 40%. In auditory-nerve fibers of goldfish saccule DGG and APV both reduced the amplitude of sound evoked EPSPs (Starr and Sewell 1991).

Interestingly, new compounds that are much more selective than KA and Q receptor antagonists have recently been developed: 6,7-dinitroquinoxaline-2,3-dione (DNQX), 6-cyano-7-nitroquinoxaline-2,3-dione (CNQX), 6,7-dichloro-3-hydroxy-2-quinoxalinecarboxylic acid (DNQC), and 3-hydroxy-2-quinoxalinecarboxylic acid (3HQC). These antagonists in dose-related manner suppressed the CAP amplitude and prolonged N_1 latency of the auditory nerve in guinea pig (Littman et al. 1989). None of the antagonists had significant effects on CM or SP. The drug concentrations causing a 50% reduction in CAP were 8 μM, 30 μM, 35 μM, and 1 μmM for DNQX, DHQC, CNQX, and 3HQC, respectively. These results are consistent with the hypothesis that KA and Q receptors are involved in neurotransmission between the hair cells and afferent nerve fibers in cochlea. Dio et al. (1990) also demonstrated that CNQX blockade of synaptic transmission from hair cells to auditory afferent nerves.

A number of criteria were formulated in order to establish a chemical as the endogenous neurotransmitter at a given synapse (see Chap. 3). The presence of high levels of L-GLU and L-ASP in both the sensory and supporting cells of the organ of Corti suggests a presynaptic location of EAA (Godfrey et al. 1976a).

A number of enzymes, e.g., glutaminase, glutamine synthetase, glutamic acid decarboxylase, and aspartate aminotransferase involved in the metabolism of L-GLU and L-ASP have been examined (Fex and Wenthold 1976; Fex et al. 1982, 1985; Wenthold 1985; Altschuler et al. 1981; Fex and Altschuler 1984; Wiet et al. 1986). However, owing to the extensive role of L-GLU metabolism in various neurons, the involvement of these enzymes in the biosynthesis of the endogenous neurotransmitter remains to be determined. It is relevant to consider here the data available on the inactivation mechanism of putative neurotransmitters in the organ of Corti. It is well established that in the nervous system the actions of amino acid transmitters are generally thought to be terminated by high-affinity Na-dependent uptake into the glia, where it is metabolized to glutamine by the enzyme glutamine synthetase. Thus, the primary afferent synapses have some means of terminating the action of the neurotransmitter. Interesting data have been obtained in studies on the uptake of a number of labeled amino acids in hair cells of the organ of Corti and the vestibular apparatus (Gulley et al. 1979; Ryan and Schwartz 1984; Schwartz and Ryan 1983; Eybalin and Pujol 1983).

The localization of glutamate and aspartate immunoreactivity has been demonstrated in both inner and outer hair cells and in spiral ganglion cells and auditory nerve fibers (Altschuler et al. 1989). Staining in inner hair cells was often more intense than in outer hair cells. Interestingly, in some hair cells the staining at the basal face was more intense.

[^{3}H]L-GLU has been shown to label preferentially cochlear efferent fibers (Gulley et al. 1979; Schwartz and Ryan 1983) or, in contrast, the inner hair cells (Eybalin and Pujol 1983). Proline, alanine, and glycine were shown to be accumulated slightly by the inner hair cells, while L-GLU, L- and D-ASP, GABA, histidine, ornithine, arginine, methionine, taurine, and choline showed no preferential accumulation in the hair cells as compared to the supporting cells of the organ of Corti in gerbils (Ryan and Schwartz 1984). Of the substances investigated, glutamine accumulated predominantly in the hair cells of the organ of Corti to a greater extent in the inner hair cells than in the outer cells. In contrast to the cochlear hair cells, no accumulation of labeled glutamine was observed in the hair cells of the vestibular apparatus. These data indicate that the hair cells of the organ of Corti possess a high-affinity uptake system for glutamine. Glutamine itself is unlikely to be a neurotransmitter; it is most likely a precursor for L-GLU synthesis, the reaction being activated by glutaminase. The existence of the high-affinity system of glutamine uptake is designed to remove the transmitter from the synaptic cleft.

Further evidence for the hypothesis that an excitatory amino acid such as L-GLU or L-ASP is the neurotransmitter released by hair cells has been published, indicating that these amino acids or structurally related substances are released into the fluid in a Ca^{2+}-dependent manner. Thus, perilymph obtained from frogs or guinea pigs subjected to sound stimulation was shown to induce an increase in firing rate in single units if infused into the perylymphatic sac of animals kept in silence (Sewell et al. 1978). However, the concentrations of both glutamate and aspartate did not differ from the levels of these amino acids in the perilymph collected in silence. It remains questionable whether the auditory nerve-activating substance is identical with or related to the endogenous transmitter. In a preliminary study, Bledsoe et al. (1981b) reported a small increase in the L-GLU concentration in the perfusate of animals exposed to sound stimulation. Medina and Drescher (1981) examined the concentration of 15 primary amines in the perilymph of guinea pigs kept in both silence and sound stimulation. No statistical noise-induced changes were found in the concentration of any compounds measured including the putative neurotransmitters, L-GLU, L-ASP, glycine, and taurine. Melamed et al. (1982) reported similar results including GABA. A detailed analysis was undertaken to study the content of 81 primary amines in the perilymph of the guinea pig cochlea (Drescher et al. 1983). Of all possible neurotransmitters, the following substances were characterized with the greatest concentrations (in mM): taurine (13.3), L-GLU (3.6), L-ASP (0.8), glycine (40), GABA (0.14), and methionine-enkephaline (4.3). Sound stimulation was accompanied by an increase in the concentration of a GABA-like component. The authors come to the conclusion that GABA seems to be an afferent neurotransmitter in the hair cells, since its concentration closely correlates with levels of sound stimulation. On the one hand, however, GABA and enkephaline seem likely to be released from efferent terminals. On the other hand, some substances released from the hair cells during sound stimulation may serve as neuromodulators. L-GLU release into the perilymph has been observed at a high level of sound stimulation (Drescher and Drescher 1985).

Extracts from the inner ears of fish have recently been shown to contain two unidentified excitatory substances: one of a low molecular weight (M about 200), and one of high molecular weight (M less than or equal to 5000) (Sewell and Mroz 1987b). Both these substances induced an increase in the discharge rate of afferent fibers innervating hair cells in the lateral line organ of *Xenopus laevis*.

An alternative technique to induce a synchronous transmitter release was used by Jenison et al. (1985) with solutions containing high K^+ concentration. The perfusion of the guinea pig cochlea with 50 mM K^+ revealed a Ca^{2+}-dependent release of L-GLU, taurine, and a coeluting taurine-GABA fraction. Although a number of interpretations are possible, the above results seem to provide evidence that presynaptic structures of the hair cells can release L-GLU and taurine in a Ca^{2+}-dependent manner. The data presented do not exclude the possibility that L-GLU and taurine could have been released from the terminals of efferent fibers or supporting cells of the cochlea. However, the source(s) of the L-GLU and taurine have yet to be established.

Calcium-dependent, potassium-evoked release of 27 primary amine-containing compounds from the trout saccular macula has been examined (Drescher et al. 1987). Taurine proved to be released from saccular macula in the highest amount, accounting for 72% of the total evoked release of primary amine-containing compounds. L-GLU and L-ASP release in a ratio of 6:1 was noted. Taurine is thought to be released from hair and supporting cells and may serve as a neuromodulator.

High K^+ induced the release of GABA, taurine, glutamate, aspartate, glycine and one of the unidentified substances from the guinea pig cochlea (Bobbin et al. 1990). It is well known that high K^+ depolarize many nervous elements and induce release not only neurotransmitters but also neuromodulators from the cells (hair or supporting) and nerve endings (afferent or efferent). Interestingly that distruction of the organ of Corti by kanamycin and ethacrynic acid significantly decreased the high K^+-induced release of GABA, aspartate, glycine. On the other hand, though the degree of release of glutamate and taurine although partly reduced nevertheless significant release still occured in treated animals.

The anthors believed that only portion of the released of glutamate and taurine observed in normal animals originated from the cells that were destroged whereas the released of GABA, aspartate and glycine most likely originated from the cells that were destroyed. Several criteria should be met before a chemical is considered a transmitter. One of these criteria is the demonstration that the chemical is released into the fluid surrounding the synapse by the depolarizing stimulus e.g., high levels of K^+ (Bobbin et al. 1991). This increased the high K^+ release of GLU and ASP and did not change the release of GABA, Tau, β-alanin. That the substances measured were not released in a Ca^{2+}-dependent manner conflicts with the findings of several previous studies that demonstrated Ca^{2+}-dependent, K^+-release of GLU and other compounds from hair cell systems. The anthors ascribe the results to an interference with either Na^+-dependent uptake processes or a Na^+/Ca^{2+} exchange carrier.

The problem of a possible involvement of ACh in synaptic transmission in the organ of Corti has long been discussed in the literature. Earlier publications indicated that ACh could function as an afferent transmitter in the organ of Corti (Gisselsson 1950). This assumption was based on the fact that acetylcholinesterase (AChE) blocking agents caused an increase in the latency of the action potentials (AP). However, more detailed investigations revealed that AChE blocking agents did not influence the latency of the AP and the CM (Gannon et al. 1966). The involvement of ACh in the synaptic transmission in hair cells of the organ of Corti was also supported by the presence of ACh in the perilymph and by an increase in AChE activity and ACh concentration in the perilymph during sound stimulation (Vinnikov and Titova 1961). Other investigators were unable to observe the release of ACh during acoustical stimulation (Gisselsson 1950; Norris and Guth 1974).

At synapses in which ACh is involved in synaptic transmission, the primary function of AChE is to catalyze the breakdown of ACh into choline and acetate. The presence of AChE in the cochlea (Churchill et al. 1956) was assumed to be usually associated with the terminals of the olivocochlear bundle rather than with afferent synapses (Iurato et al. 1971). Thus, after section of the crossed and uncrossed olivocochlear bundles, the enzyme activity associated with ACh synthesis (acetylcholinetransferase) was much reduced or eliminated (Jasser and Guth 1973). This indicates that the structures which remained after degeneration, including hair cells, were unable to synthesize ACh, and that the presence of the enzyme was associated with the efferent synapses which represent the major control system for the motile events of outer hair cells through ACh receptors (Plinkert et al. 1990). This conclusion is supported by the analysis of the distribution of acetylcholinetransferase and AChE in the endings of efferent nerve fibers (Godfrey et al. 1976b; 1984; Altschuler et al. 1985; Wiet et al. 1986). AChE is well known to exist in several molecular forms in both neural and non-neural tissues, and has been also implicated in the breakdown of the neuropeptides enkephalin and substance P. Of six molecular forms of AChE recognized in all vertebrate species, four were observed recently in the cochlea (Emmerling and Sobkowicz 1988).

It is noteworthy that Flock and Lam (1974) demonstrated by means of labeled precursors of ACh synthesis (methyl-[^{3}H]-choline chloride) that those organs, i.e., toadfish lateral organ and the crista ampullaris of the skate, which have efferent innervation were capable of synthesizing radiolabeled ACh. On the contrary, the frog basilar papilla, with no efferent innervation, was not able to synthesize ACh. Hence, biochemical evidence supports the hypothesis that the efferent fibers are cholinergic.

In electrophysiological experiments on ACh effects on the hair cells of the organ of Corti, this substance at a concentration of 50 μM was shown to cause an increase in the CM amplitude and a corresponding decrease in the CAP of the auditory nerve (Bobbin and Konishi 1974). Both of these processes resemble the effects of crossed olivocochlear bundle stimulation and are blocked by curare (Robertson and Johnstone 1978).

ACh was found to have no influence on the CM and CAP when administered intracochlearly in the guinea pig (Bobbin and Konishi 1974). Moreover, Galley et al. (1973) demonstrated that cholinergic blocking agents (hexamethonium, curare, and atropine) in the cat abolished the crossed olivocochlear bundle effect on CAP and CM.

Electrophysiological experiments on the organ of Corti indicated that ACh in a concentration of 50 μM caused a reduction in the background and evoked activity of single fibers of the auditory nerve (Comis and Leng 1979). Thus, there is strong evidence that ACh is not an afferent neurotransmitter in the hair cells of the organ of Corti. There is ample evidence that ACh is an efferent transmitter (Robertson and Johnstone 1978; Klinke 1981).

A possible role of catecholamines as afferent neurotransmitters in the organ of Corti has been investigated in a number of studies. In experiments on single fibers of the cat's auditory nerve (Bobbin and Thompson 1978; Klinke and Evans 1977) and guinea pigs (Comis and Leng 1979), the perfusion of the scala tympani with catecholamines at a concentration of 10^{-4} to 10^{-5} M was found to reduce the resting firing rate or produce no effect at all. The CAP and CM amplitudes of the guinea pig experienced no changes under the influence of catecholamines (Bobbin and Thompson 1978); α- and β-adrenoreceptor blocking agents (phentolamine, propranolol) had no effect on CAP and afferent discharges of the cochlear nerve (Klinke and Evans 1977).

Histochemical fluorescent methods failed to reveal the presence of catecholamines in the afferent synapses of hair cells (Hilding 1964; Fex et al. 1965; Spoendlin and Lichtensteiger 1966; Densert 1974). Treatment of afferent synapses with 6-hydroxydopamin, which is known to produce a depletion of adrenergic vesicles, did not alter the appearance of the afferent synapse (Borg et al. 1974; Densert 1974). Thus, catecholamines do not seem to be afferent neurotransmitters in the hair cells of the cochlea.

Numerous investigations have been devoted to the study of GABA in synaptic transmission in the cochlea. Available data on the role of GABA in the afferent transmission in the organ of Corti are contradictory. Thus, biochemical methods failed to detect GABA in the inner and outer hair cells of the cochlea (Tachibana and Kuriyama 1974; Godfrey et al. 1976a,b). It is worth noting that antibodies to GABA showed no immunoreactivity in guinea pig hair cells or spiral ganglion cells (Thompson et al. 1986; Altschuler et al. 1989; Eybalin et al. 1988; Usami et al. 1988). Moreover, GABA synthesis in the cochlear receptors was found to be associated with efferent rather than afferent synapses (Schwartz and Ryan 1983). The activity of glutamic acid decarboxylase (GAD) was low in the inner and outer hair cell region of the cochlea (Fex and Wenthold 1976). This evidence opposes the idea that GABA is a likely afferent transmitter in the organ of Corti. Labeled GABA uptake is mainly noticed on the terminals of efferent nerve fibers (Gulley et al. 1979; Schwartz and Ryan 1983).

The activation of the guinea pig cochlear receptors induced a significant increase in the GABA-like component concentration in the perilymph (Drescher et al. 1983). The authors believed that a GABA-like component may mediate excitatory receptonearal transmission in the cochlea. However, GABA is likely to be released from efferent terminals as well. On the other hand, GABA accumulation can be associated with decarboxylation of the released GLU, since the presence of small GAD activity has been demonstrated in all neural tissues. The guinea pig cochlea perfusion with 50 mM K^+ revealed a Ca^{2+}-dependent release of a coeluting taurine-GABA fraction.

Electrophysiological investigations on the effect of GABA and its antagonists on the synaptic transmission in the hair cells of the cochlea are not consistent with the view that GABA is an afferent neurotransmitter. Intracochlear infusion of 10 mM GABA and 1 mM bicuculine had no effect on the CAP and CM of the guinea-pig cochlea (Bobbin and Thompson 1978). Picrotoxin, a GABA antagonist, did not block the evoked activity or affect the CAP amplitude of the cat auditory nerve (Desmendt and Monaco 1962; Bobbin and Guth 1970; Klinke and Oertel 1977b). Thiosemicarbazide, an agent capable of inhibiting GAD and thus GABA synthesis, caused no change in the CAP parameters of cat auditory nerve (Desmendt and Monaco 1962; Bobbin and Guth 1970).

Histochemical methods did not reveal the presence of histamine in the organ of Corti (Spoendlin and Lichtensteiger 1966). This substance, in a very high concentration (10 mM), reduced the CAP amplitude in the cat auditory nerve by 19% (Bobbin and Thompson 1978).

Histochemical and biochemical methods failed to detect synthesis or the presence of serotonin in the organ of Corti (Fex et al. 1965; Borg et al. 1974; Densert 1974). Serotonin antagonist injection into the cochlea had no effect on the electrical activity in the nerve fibers of the auditory nerve (Klinke and Oertel 1977a). At the same time, serotonin reduced the CAP amplitude in the auditory nerve of the guinea pig at a rather high concentration (10 mM) (Bobbin and Thompson 1978). On the other hand, serotonin did not affect the electrical activity of cat cochlea (Klinke and Oertel 1977c).

Taurine in a concentration up to 10 mM had no effect on the CAP amplitude of the auditory nerve or the CM of the guinea pig (Bobbin and Thompson 1978).

Substance P produced no effect on the electric parameters in guinea-pig cochlea (Norris et al. 1976). Adenosine triphosphate (ATP) at concentrations of 1 mM and 10 mM reduced the CAP in the guinea-pig auditory nerve by 38 and 95%, respectively (Bobbin and Thompson 1978). The authors attributed these effects to the ability of ATP to chelate calcium.

Octopamine in a concentration of 10 mM had no effect, while tyramine reduced the CAP amplitude in the guinea-pig auditory nerve (Bobbin and Thompson 1978). The authors believed that tyramine can be regarded as a putative neurotransmitter in the cochlea. However, so far there have been no experimental data supporting this point of view.

It has been recently demonstrated by biochemical analysis that hair cells from the trout saccular macula contain large amounts of a histidine-containing compound which has been identified as N-acetylhistidine (Drescher and Drescher 1991). The anthors believe that N-acetylhistidine may have a neurotransmitter role.

To date, a number of investigators have suggested the idea of a possible involvement of neuropeptides in synaptic transmission in the receptors of the acousticolateral system, but no precise experimental evidence has been offered so far (Wenthold 1985; Hoffman 1986). Such investigations were carried out on receptors in the organ of Corti. Thus, the vasoactive intestinal peptide was found in the perilymph (Lim et al. 1981). Using immunohistochemical methods, the presence of enkephaline-like immunoreactivity was first shown in efferent fiber endings in the cochlea of the cat and the guinea pig (Fex and Altschuler 1981). It should be noted, however, that no immunofluorescence was found in the spiral ganglion cells in the auditory nerve fibers or in the hair cells.

Nevertheless, using the high-performance method of liquid chromatography, it was possible to isolate two enkephaline-like peptides from the guinea-pig cochlea (Hoffman et al. 1983). Moreover, methionine-enkephaline-like peptide concentration in the perilymph of the guinea-pig inner ear was shown to become elevated following acoustic stimulation (Drescher et al. 1983). The authors suggested that enkephaline may have been released from the efferent terminals. It is worth noting that noise exposure lowers the endogenous levels of methionine-enkephaline in the guinea pig cochlea (Eybalin et al. 1987b).

There is evidence for the presence of several biologically active opioid peptides such as enkephaline, dynorphin B, and alpha neoendorphin, and opiate receptors in the cochlea (Eybalin et at. 1983, 1984, 1985a,b, 1987a; Altschuler et al. 1984; Eybalin and Pujol 1984; Hoffman et al. 1984, 1985; Lehtosalo et al. 1984; Gil-Loyzaga et al. 1988). At present, there is little evidence that opioid peptides serve as neurotransmitters in the organ of Corti. Recent studies have demonstrated the neuromodulatory role of opiates. Kuznetsov and Godukhin (1985) have revealed that methionine-enkephalin reduces GLU binding to synaptic membranes by decreasing the number of binding sites without affecting the binding affinity.

It is of interest that opioid receptors in the brain are linked to the cyclic AMP system. The presence of the μ- and δ-cochlear opioid receptors has been recently shown (Eybalin et al. 1987). The same authors have demonstrated that opiate agonists reduce the synthesis of cyclic AMP in the cochlea. These receptors may be located on both the pre- and postsynaptic side of the afferent synapse and could modulate the release of a neurotrnsmitter or postsynaptic responses.

Calcitonin gene-related peptides (CGRP) have been shown to be found in the lateral system of efferents in the organ of Corti in guinea pig (Takeda et al. 1987; Sliwinska-Kowalska et al. 1989), in rat (Kitajiri et al. 1985; Takeda et al. 1986), in rat and cat (Lu et al. 1987). A few CGRP-immunoreactive structures were found in the region of outer hair cells in rat (Kitajiri et al. 1985). The functional role of the presence of CGRP is not known so far. Yolikoski and Lehtosalo (1985) suggested that in the human cochlea neuropeptides can play a role in the temporary threshold shift after noise exposure.

4.2 The Vestibular Apparatus

EAA and their agonists have been considered as putative neurotransmitters in vestibular organs (Dechesne et al. 1984). Valli et al. (1985) observed the excitatory effect of L-GLU on the receptors of the frog semicircular canal. Slow ampular potentials reflecting the electrical activity of the sensory epithelium in the hair cells experienced no changes in L-GLU solution. On the contrary, L-GLU induced an increase in the firing frequency of afferent fibers. It is noteworthy that L-GLU did not influence the hair cell responses to mechanical stimuli when added to the endolymphatic fluid. However, their experiments failed to demonstrate the restoration of impulse activity in the afferent fibers upon perfusion with L-GLU when endogenous neurotransmitter release was blocked by low Ca^{2+}-high Mg^{2+} solution. Under these conditions L-GLU caused only a long-lasting depolarization in the afferent fibers which was not accompanied by impulse activity. In this relation, the exogenous L-GLU has been suggested to have at least two sites of action, a presynaptic and a postsynaptic one. The presynaptic action is responsible for an increased release of the natural transmitter which in turn produces a substantial increase in the firing rate in afferent fibers. This L-GLU effect was antagonized by DLAA. On the other hand, L-GLU produced a depolarization of the postsynaptic membrane of afferent fibers which caused no afferent discharges by itself.

A more detailed analysis of the effects of EAA agonists in the frog vestibular system have shown that Q, KA, and NMDA exerted a complex excitatory action consisting of a rapid and brief increase in frequency of both EPSPs and spikes followed by longer-lasting depolarization of afferent fibers, Q and KA being more potent than NMDA (Prigioni et al. 1990). High Mg^{2+} solution caused a decrease of the agonist-induced afferent discharge, without substantially affecting axonal depolarization. These results are consistent with the hypothesis that EAA agonists act both pre- and postsynaptically. The authors believed that presynaptic amino acid receptors may function as autoreceptors controlling the afferent transmitter release by a positive feed-back mechanism, and thereby may increase the gain of the mechanoelectrical transduction in the hair cells.

Studies on the frog vestibular apparatus by other authors were devoted to the investigation of EAA agonists (Annoni et al. 1984). L-GLU and L-ASP increased the frequency of background activity in afferent fibers (Fig. 66A). The threshold for activation was 0.1–0.2 mM. NMDA was relatively more potent, while KA and Q were considerably more potent in their ability to activate the afferent fibers (Fig. 66B). In contrast to the findings of Valli et al. (1985), the effects of L-GLU and L-ASP on the firing rate of afferent fibers persisted in high-Mg^{2+} solution. The authors believed that these amino acids and their agonists induced a reversible depolarization of vestibular afferents. The elevation in the frequency of discharges in the

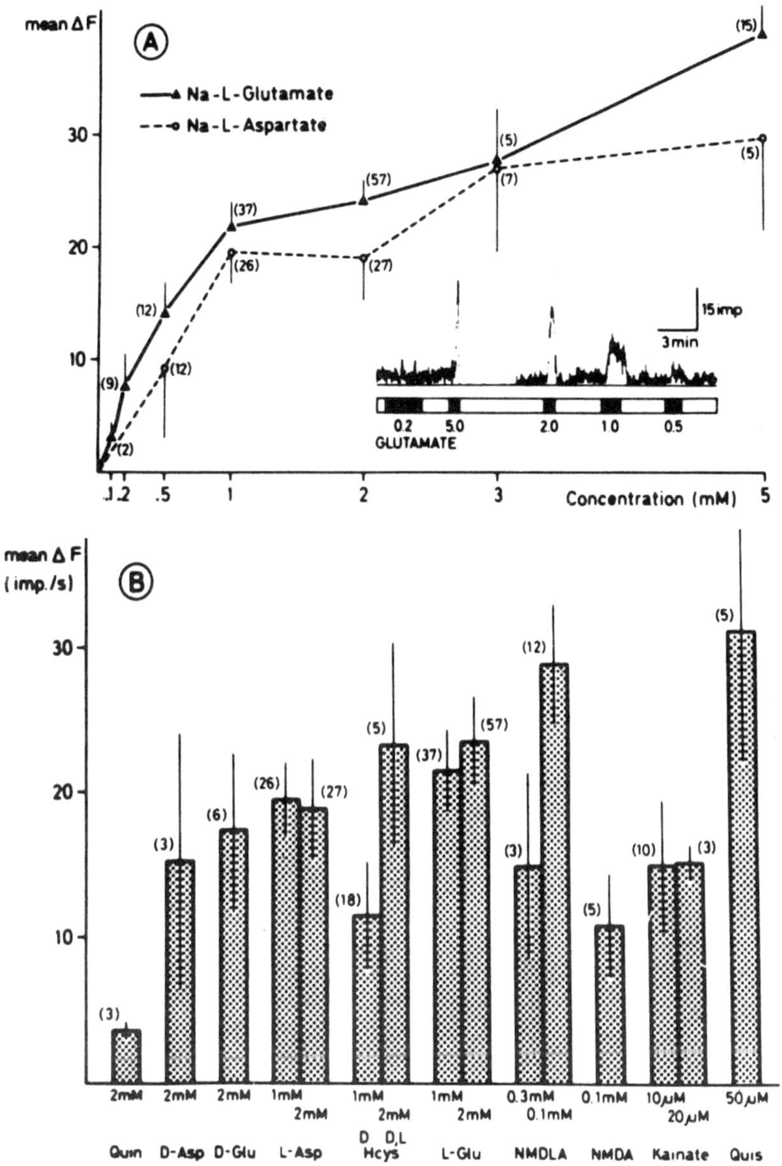

Fig. 66A,B. Dose-response characteristics of acidic amino acids on afferent excitation in the vestibular hair cell-afferent synapse of the frog recorded extracellulary. **A** Dose-response curves to glutamate (*solid triangles*) and aspartate (*open circles*). The *numbers in parentheses* indicate the number of afferents tested at a given concentration. *Vertical bars* represent SEM. *Inset* Responses of one cell to a given concentration of glutamate (in millimolar concentration). In this case, 5 mM glutamate produced depolarization inactivation of the afferent. **B** Responses of afferents to a given concentration of acidic amino acids. The *numbers in parentheses* represent the number of afferents tested at a given concentration. *Vertical bars* represent SEM. ΔF represents the peak frequency obtained during drug application minus the resting firing frequency of the axon prior to drug administration. Hcys homocysteic acid; Glu glutamate; *Asp* aspartate; *NMDA* N-methyl-D-aspartic acid; *Quin* quinolinic acid; *kainate* kainic acid; *Quis* quisqualic acid. (Annoni et al. 1984)

afferent fibers persisted throughout the whole period of agonist applications, demonstrating the lack of postsynaptic receptor desensitization. EAA and their agonists in concentration of 10^{-4} up to 10^{-6} M exerted an increase in the firing rate of primary afferents in the vestibular system of the axolotl, *Ambystoma mexicanum* (Soto and Vega 1988). The rank order in potencies of agonists was KA>Q>D-ASP>L-GLU>L-ASP>NMDA. It is worth noting that the EAA and the influence of its agonists experienced no significant changes in high-Mg^{2+} and low-Ca^{2+} solution, indicating their postsynaptic action.

A wide variety of antagonists of postsynaptic amino acid receptors blocked synaptic transmission and postsynaptic amino acid-induced depolarizations (Annoni et al. 1984; Soto and Vega 1988). Among the antagonists, the most effective ones were KENYA, APV, APB, DGG, and PDA. GDEE and DLAA were less potent. Both APV and KENYA decreased the amplitude of the spontaneous postsynaptic potential without affecting its frequency. The antagonists blocked the EAA, and their agonists evoked depolarization of the postsynaptic membrane. The effects evoked by L-GLU were found to be more resistant to antagonists than to L-ASP, Q, and KA. At a concentration of 1 mM, APV completely blocked the L-ASP-induced postsynaptic depolarization. The excitatory effects of KA and L-ASP were blocked by KENYA.

The data available on vestibular hair cell synaptic transmission suggest that the most probable candidates are L-GLU and L-ASP (Guth et al. 1988). As to the nature of postsynaptic receptors, the existance of several specific types of receptor or single type of receptors with nonspecific sensitivity to the known antagonists is postulated. It is noteworthy that in high-Mg^{2+} solution there was no reduction either in the postsynaptic potential amplitude or in the amino acid-induced excitation. Hence, Mg^{2+} effect on postsynaptic receptors and NMDA-induced depolarization seem quite unlikely (Annoni et al. 1984).

It has been recently demonstrated that streptomycin blocked the postsynaptic excitatory action produced by KA and Q in primary afferents of the vestibular organ of axolotl (Pérez et al. 1991) and synaptic transmission in receptors of the semicircular canals of the frog (Zucca et al. 1992).

Biochemical studies also provide evidence that L-GLU acts as neurotransmitter in vestibular hair cells of the cat (Raymond et al. 1984). In sensory hair cells of the vestibular system of mouse, rat, and cat, GLU, apart from its metabolic function, may play a role in synaptic transmission, since GLU-like immunoreactivity was found in sensory cells whereas the supporting cells were immunonegative (Dememes et al. 1990).

Calcium-depended release of L-GLU from frog vestibular organs in a 5 mM K^+ solution was demonstrated by means of a bioluminescence-enzymatic method (Zucca et al. 1992).

The results concerning a possible role of GABA in the synaptic transmission in the hair cells in the vestibular system are not definite. In earlier investigations, an excitatory effect of GABA on the receptors of the cat macula sacculi was observed (Felix and Ehrenberger 1977, 1982). Flock and Lam (1974) first suggested that GABA might be an afferent neurotransmitter in the labyrinthine sensory organ. They showed that bullfrog basilar papilla with only afferent innervation, and the crista ampularis of the semicircular canal of the skate with both afferent and efferent innervation were capable of synthesizing GABA from radiolabeled glutamate. Moreover, they demonstrated the presence of GAD in the basilar papilla. The authors also found that picrotoxin blocked both resting and evoked activity in the afferent nerve fibers of the crista ampullaris of the skate. Chick vestibular sensory epithelium has been shown to be capable of binding and synthesizing GABA from glutamate (Meza 1984, 1985; Meza et al. 1982, 1985). GABA-like immunoreactivity in the chick vestibular organ was confined to the cytoplasm of the hair cells (Usami et al. 1987).

In experiments with the hair cells of the vestibular system of the bullfrog, GABA and picrotoxin had no effect on the impulse activity in afferent fibers (Guth and Norris 1984). In the afferent fibers of vestibular system of axolote, the GABA and muscimol produced a very weak excitatory effect (Vega et al. 1987). Moreover, the GABA antagonists bicuculine and picrotoxin had no significant effect on these fibers. In frog vestibular receptors, GABA and its potent agonist, muscimol, occasionally excited the afferent fibers (Annoni et al. 1984). However, when the synaptic transmission was blocked with high-Mg^{2+} and low-Ca^{2+} solution, neither GABA nor muscimol had a significant effect upon the membrane potential of the afferent fibers. It is noteworthy that in these conditions the L-GLU and L-ASP evoked postsynaptic effect. In addition, bicuculine and picrotoxine exerted no significant effect upon spontaneous synaptic potentials (Annoni et al. 1984). It should be emphasized that calcium-dependent, potassium-evoked GABA release from the trout saccular macula was also observed (Drescher et al. 1987).

In conclusion, so far the bulk of the data obtained makes it impossible to consider GABA as an afferent neurotransmitter in the vestibular hair cells.

The question of possible ACh involvement in synaptic transmission in the receptors of vestibular organ has long been discussed. Flock and Lam (1974) demonstrated by means of labeled precursors of ACh synthesis that those organs (the crista ampullaris of the skate) with efferent innervation were capable of synthesizing ACh. In contrast, the bullfrog basilar papilla, with no efferent innervation, was not able to synthesize ACh. Hence, the biochemical findings support the hypothesis, suggesting that ACh could well be a neuromediator in efferent synapses (Meza 1985).

In electrophysiological experiments, carbachol was shown in some cases to excite the afferent nerve fibers of vestibular hair cells in the frog (Annoni et al. 1984). This excitation appears to result from an increase in synaptic potential frequency rather that from a direct depolarization of the afferent fibers. A detailed investigation studying an effect of cholinergic agonists upon afferent transmission in the vestibular labyrinth of the frog revealed an increase in the firing rate of afferent fibers induced by both muscarinic and nicotinic agonists (Bernard et al. 1985). These effects turned out to be blocked by prior atropine or curare administration. The cholinergic agonists alter a frequency of spontaneous synaptic potential, indicating a presynaptic action upon the hair cell. This conclusion is supported by the fact that cholinergic agonists did not influence the resting membrane potential of the vestibular afferent when the transmitter release had been blocked by perfusion with high-Mg^{2+}/low-Ca^{2+} solution. Moreover, curare did not block the mechanically evoked responses in the isolated frog labyrinth when applied to the synaptic region (Rossi et al. 1980). On the other hand, when d-tubocurarine was applied to the isolated open semicircular canal, it was capable of depressing or abolishing both the resting discharge of the ampular receptors and their responses to mechanical stimulation (Valli et al. 1974). D-tubocurarine seems likely to interfere with transduction processes at the apical membrane.

Norris et al. (1988) and Housley et al. (1990) postulated the existence of at least three types of cholinergic receptors in the semicircular canal in the frog. Here a facilitatory receptor is activated by muscarinic and nicotinic agonists and blocked by atropine (10 μM), and partially by curare (50 μM). Two other types of receptors produce suppression of the vestibular afferent firing rate. They are blocked by strychnine and curare, respectively. The facilitatory effect of ACh is likely to be produced by its influence on of the hair cell activity, inducing the release of the hair cell transmitter and thus the firing of afferent fibers. The site of the "curare-preferring" receptors may be located on either pre- or postsynaptic membrane of the efferent synapse. The strychnine-sensitive suppressive effect of ACh may also be associated with the efferent synapse. Thus, ACh is unlikely to be an afferent neurotransmitter in receptors of the vestibular system.

A possible role of catecholamines, dopamine, serotonin, and histamin in the receptors of the vestibular system has been studied in a number of investigations. In morphological studies, a number of authors detected changes in diameter and density of the hair cell synaptic bars of the frog vestibular epithelium following treatment with reserpine or monoamine oxidase (Thornhill 1972; Osborne and Thornhill 1972). These results seem to support the idea of possible catecholamine involvement in synaptic transmission. However, experiments with radioactive precursors failed to detect the synthesis of dopamine, noradrenaline, or serotonin in bullfrog basilar papilla and semicircular canal in skate (Flock and Lam 1974). Adrenaline has been shown in some cases to excite the afferent fibers of vestibular organ in frog (Annoni et al. 1984).

Interestingly, some substances may be implicated in both neurotransmission and neuromodulation. For instance, in the semicircular canal of the frog, adenosine is thought to be a neuromodulator of endogenous transmitter release (Bryant et al. 1987). Adenosine was shown to inhibit the firing rate of afferent fibers. This effect was antagonized by theophylline, which increased the firing rate. On the other hand, histamine was found to have a facilitatory effect in the same preparation (Housley et al. 1988). This effect of histamine was antagonized by an inhibitor of its synthesis, fluoromethylhistidine, in dose-dependent manner. Histamine antagonists were shown to suppress the firing rate. These results suggest that histamine may have some neuromodulatory role in the vestibular receptors. A similar role has been suggested for some neuropeptides, particularly enkephalines (Soto et al. 1988). Naloxone has been established as inducing an increase in the resting firing rate of afferent fibers of the vestibular system of the axolotl. Nevertheless, naloxone modifies neither the response of afferent fibers to mechanical stimulation nor the effect evoked by L-GLU or KA.

4.3 Lateral-Line Organ

Mechanoreceptors of the lateral-line organs in fish and amphibians are also activated by L-GLU and L-ASP. Thus, the background activity of nerve fibers associated with the neuromast of the catfish lateral line increased upon superfusion with solution containing L-GLU in a very high concentration up to 50 mM (Katsuki 1973). L-GLU and L-ASP in a concentration of 10^{-2} to 10^{-3} M have been shown to excite highly afferent nerve fibers of the neuromasts of *Xenopus laevis* (Russell 1968; Zimmerman 1979; Bobbin et al. 1981b, 1983, 1985; Bledsoe et al. 1983 et al.; Mroz and Sewell 1989). At a high concentration, the excitatory phase is rapidly followed by a suppression of spontaneous activity suggestive of a depolarization blockade (Fig. 67). The effect induced by L-ASP was somewhat stranger than that of L-GLU, while the D-GLU potency remained about half that of the L-form. KA exerted an excitatory effect at concentration of 5–20 μM, i.e., it was about 250 times as effective as L-GLU. Dihydrokainic acid had no effect in concentrations of up to 10 mM. Q was the most potent agonist, increasing the resting activity at concentration of as low as 10^{-6} M and was about 10^3 times more potent than L-GLU. Other glutamate analogs, L- and D-homocysteate in equal potency, also induced a strong excitatory effect on the lateral-line mechanoreceptors; the L-form was about ten times as effective as L-GLU. Unlike the cochlea, however, the lateral-line mechanoreceptors are quite responsive to NMDA, similar to those of L-GLU. Hence, in the lateral-line organ the postsynaptic amino acid receptors include those of the NMDA-preferring subtype.

One of the most important approaches for identifying the transmitter substance is the use of specific antagonists blocking the responses of afferent fibers not only to natural stimulation but also to the application of the putative neurotransmitters and their agonists. An important argument in favor of the existing NMDA receptors in the sensory synapse of the lateral-

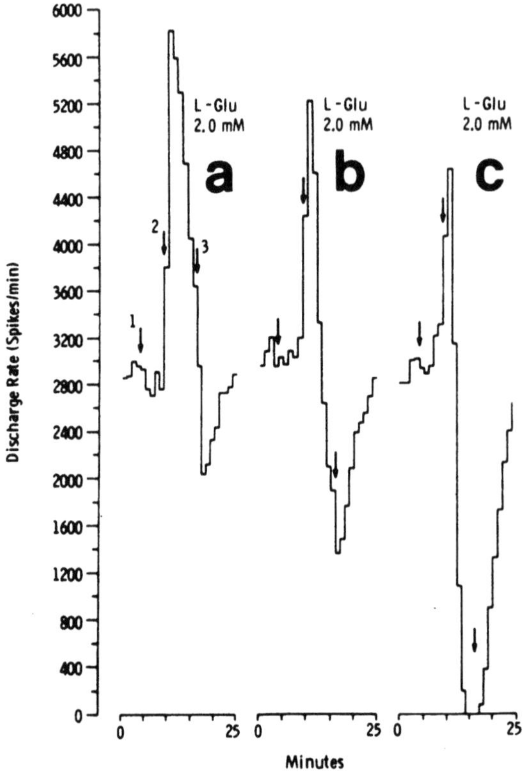

Fig. 67a–c. Three lateral-line afferent nerve fiber responses elicited by 2.0 mM L-glutamate (L-GLU) illustrating changes in response characteristics to repeated application at the same concentration. The *three arrows* (*1,2,3*) in each response designate the time of: *1* Ringer wash off; *2* drug application; *3* Ringer wash on. Note that the magnitude and duration of the excitation decrease, while those of the post-excitatory suppression increase for the second (**b**) and third (**c**) applications. (Bledsoe et al. 1983) (Reprinted with permission from Comp. Biochem. Physiol. v. 75C, Bledsoe et al. Comparative actions of glutamate and related substances on the lateral line of Xenopus laevis. Copyright 1983, Pergamon Press PLC.)

line organs is supported by the effect induced by DAA, an NMDA receptor antagonist. DAA, at a concentration of 0.25–0.5 mM, reduced resting and evoked activity of the *Xenopus* lateral-line sensory organs, while the L-optic isomer was inactive (Bledsoe and Bobbin 1982). The responses induced by NMDA (1–2 mM) have been shown to be most sensitive to DAA (0.25–0.5 mM), whereas those evoked by L-ASP and L-GLU (1–2 mM) were less sensitive, and the responses to KA were least affected by DAA. Since the DAA concentration was essential for the suppression of both the evoked and NMDA-induced responses, NMDA, DAA, and the endogenous neurotransmitter may be assumed to act upon the same postsynaptic receptors. Moreover, these results strongly support the notion that the afferent transmitter in the lateral-line receptors is an excitatory amino acid acting, at least in part, on NMDA receptors.

There are data on the effect of several excitatory amino acid antagonists on synaptic transmission in the *Xenopus* lateral line including APV, GDEE, PDA, DGG, and KENYA (Bledsoe et al. 1985). GDEE has been shown to block no spontaneous activity in the *Xenopus* lateral-line organ (Russell 1976). On the other hand, PDA and DGG are the most potent antagonists, suppressing both resting and evoked activity at a concentration as low as 50–100 μM. KENYA is less potent (125–250 μM) (Bledsoe et al. 1985). Interestingly, at a concentration of 1–2 mM, PDA and DGG blocked the evoked activity totally but only partially suppressed the resting activity. On the other hand, APV (250 μM) blocked the NMDA-evoked responses and suppressed resting but not evoked activity. These data suggest that, in addition to NMDA receptors, KA and Q receptors are also involved in afferent transmission in lateral-line receptors. The functional role of these receptors remains to be established.

The effects of 56 compounds on the discharge rate of afferent fibers innervating hair cells have recently been examined in the lateral-line organ of *Xenopus laevis* (Mroz and Sewell 1989). These compounds included amino acids, glutamyl dipeptides, and neurotransmitter candidates. Of these compounds, the neutral (alanine, serine, threonine, asparagine, glutamine, and proline) and acidic amino acids (L-GLU, L-ASP, and cysteine sulfinate) produced an excitatory effect.

An important criterion in the identification of a transmitter is its release upon natural stimulation of the sensory organ. One of the studies dealt with the release of endogenous GLU and ASP from the isolated skin of *Xenopus laevis* with and without lateral-line stitches (Bledsoe et al. 1980). The skin containing stitches was found to have much greater GLU and ASP concentrations than that containing no receptors. Subsequent mechanical stimulation caused a significant increase in a GLU efflux. The results were less clearcut with respect to the ASP efflux.

An alternative technique of afferent neurotransmitter release stimulation was an in vitro superfusion technique with a K^+-rich solution and high performance liquid chromatography (Bledsoe et al. 1989a). Using this technique, Ca^{2+}-dependent release of GLU, ASP, and two as yet unidentified substances from tissues containing the lateral-line organ was demonstrated. The small and variable release of GLU and ASP from nonstitch tissue was also demonstrated. It is thus, conceivable that GLU and ASP are released from a number of sites including structures of lateral-line neuromasts (supporting cells, afferent and efferent terminals) and structures of epidermal layer (epithelial cells, small glands). In summary, the available data are consistent with the hypothesis that L-GLU or its structural analogs serve as transmitters in the lateral-line receptors.

There is considerable evidence indicating that ACh is an efferent transmitter in receptors of the acousticolateral system. ACh and its agonists have been shown to suppress resting afferent activity (Russell 1971). Flock and Lam (1974) demonstrated that lateral-line canal organ in the toadfish synthesized ACh and contained the choline acetyltransferase.

In the mechanoreceptors of the *Xenopus laevis* lateral-line organs, ACh (1.25–80 μM) and carbachol (1.25–40 μM) caused a dose-dependent increase in resting activity (Winbery et al. 1983; Bobbin et al. 1985). Atropine, the ACh receptor antagonist, blocked the excitatory effects of ACh and carbachol, reduced the frequency of resting firing rate in the nerve fibers, but had no effect on water motion-induced activity. Responses to ACh and carbachol were reduced by the high Mg^{2+}/low-Ca^{2+} solution. The excitatory effect produced by ACh and carbachol appears to be due to presynaptic action on the hair cells, causing a release of endogenous neurotransmitter.

The role of GABA in the afferent transmitter in lateral-line receptors has been discussed in the literature. Flock and Lam (1974) demonstrated GABA but not dopamine, noradrenaline, or serotonine synthesis in the lateral canal organ, which has both afferent and efferent innervation. This organ contains glutamate decarboxylase. At a concentration of 10^{-3}–10^{-7} M, GABA and its antagonists picrotoxin and bicuculin had no influence on the spontaneous activity in frog neuromasts (Russell 1976). On the other hand, GABA (0.019–1.25 mM) suppressed spontaneous activity of the afferent nerve fibers of the *Xenopus* lateral-line organ (Bledsoe et al. 1983; Bobbin et al. 1985). This GABA influence was blocked by bicuculine (100 μM). It is worth noting that bicuculine (100 μM) had no effect on water motion-evoked activity. On the other hand, GABA produced a very brief excitation followed by suppression of the afferent discharge (Mroz and Sewell 1989). These effects appeared to be specific for GABA, as several structural GABA analogs were all inactive. So far the bulk of the data obtained makes it impossible to consider GABA as a neurotransmitter in the lateral-line receptors.

Adrenaline and noradrenaline had no influence on spontaneous impulse activity in the afferent nerve fibers of *Xenopus* lateral-line receptors (Russell 1976).

A detailed analysis of the effect of histamine on the lateral-line receptors in *Xenopus* indicates that this neurotransmitter (0.002–2.0 mM) had no effect on resting and evoked activity (Bledsoe et al. 1989b, Mroz and Sewell 1989). In contrast, pyrilamine, an H_1 receptor antagonist, and cimetidine, an H_2 receptor antagonist, both suppressed resting activity in the afferent fibers. These effects of histamine receptor antagonists were nonspecific, since pyrilamine suppressed nonselectively the excitatory effect of L-GLU, L-ASP, KA, Q, and NMDA. Hence, these results do not support the notion that histamine serves as a neurotransmitter or modulator in *Xenopus* lateral-line afferent synapse. ATP in 0.5–5.0 mM concentration, produced an excitatory effect on these receptors (Mroz and Sewell 1989).

It is worth noting that several glutamyl dipeptides produced a concentration-dependent suppression of discharges (Mroz and Sewell 1989; Sewell and

Starr 1991). On the other hand, CGRP at a concentration of 10^{-4} to 10^{-5} M increased the discharge rates of afferent nerve fibers of the lateral-line organ of *Xenopus laevis* (Adams et al. 1987). Although other interpretations are possible, the authors believed that CGRP may serve as an excitatory efferent neurotransmitter since CGRP-like immunoreactivity is localized in small, presumably efferent, fibers. This notion is consistent with the observation of a noncholinergic excitatory effect produced by efferent stimulation in the lateral-line organ (Flock and Russell 1973; Rossi et al. 1980).

4.4 Electroreceptors

During the past decade, great achievements have been made in investigation of the nature of synaptic transmission in electroreceptors (Akoev and Andrianov 1989). However, the afferent transmitter in the electroreceptors still remains to be identified. The most probable candidates are the excitatory amino acids, L-GLU and L-ASP. Indeed, L-GLU (10^{-6} M) increased the firing rate of afferent nerve fibers and decreased the thresholds to electrical stimulation of electroreceptors in mormyrid fish (Steinbach and Bennett 1971).

In experiments on the ampullary electroreceptors of the transparent catfish, *Kryptopterus*, L-GLU (10^{-6} M) was shown to increase the resting firing rate of the afferent fibers and the responses of receptors to electrical stimulation, while D-GLU and L-ASP had a little effect (Teeter and Bennett 1981) (Fig. 68). The latter effect may have been due to a summation of the actions of the endogenous neurotransmitter and the exogenous L-GLU. At high concentration (10^{-3} M) L-GLU had a strong transitory excitatory effect followed by depression of activity, apparently due to an excessive depolarization of the afferent fiber membrane since the fiber became unexcitable to direct stimulation. In this case, the evoked activity was also blocked. It is interesting to note that L-GLU restored the receptor resting activity blocked by high Mg^{2+} solution. Thus, it is very probable that L-GLU does act on the postsynaptic membrane receptors. The authors believe that the desensitization of postsynaptic receptors was not prominent in tonic receptors in the face of a continuous transmitter release from the sensory cells.

In vitro preparation of ampullary electroreceptor organs of the teleost fresh-water catfish Ictalurus nebulosus was shown to be promising model system for neuropharmacological research of synaptic transmission (Andrianov et al. 1992a). Behavior of the organs recorded in vitro was much the same as in vivo. Both the mucosal and the serosal side of the skin could be exposed to a variety of drugs. L-GLU (0.1–1 mM), Q (0.8–40 μM) and KA (10–250 μM) applied externally to the serosal side of the skin strongly increased both resting and electrically-evoked activity in the afferent fibres. NMDA had no effect, even at a concentration of 5.0 mM (Andrianov et al.

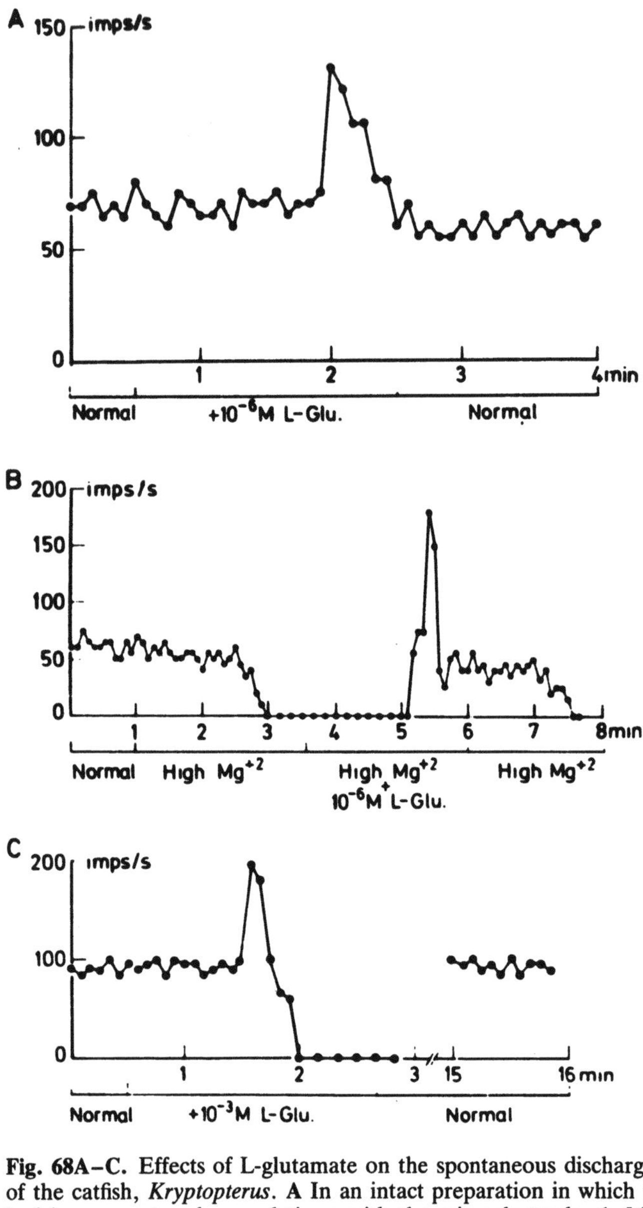

Fig. 68A–C. Effects of L-glutamate on the spontaneous discharge from a tonic receptor of the catfish, *Kryptopterus*. **A** In an intact preparation in which the receptor epithelium had been punctured several times with the microelectrode, 1 μM L-glutamate evoked a transient acceleration in the resting discharge, followed by a decrease to below the original resting level. Spontaneous activity returned to its original level when the preparation was washed with normal saline. **B** In an isolated fin preparation, saline containing 15 mM Mg^{2+} and 0 mM Ca^{2+} blocked the spontaneous discharge. Addition of 1 μM L-glutamate to the Mg^{2+} saline resulted in a transient acceleration in firing which rapidly decreased to a fairly steady discharge, slightly slower than the original resting level. Return to high Mg^{2+} saline containing no glutamate resulted in a complete cessation of activity. **C** In an isolated fin preparation 1 mM L-glutamate evoked a transient increase in the spontaneous discharge rate which was followed by a complete suppresion of firing. Washing with normal saline resulted in a return of the spontaneous discharge to about its original level. Frequencies were measured as five times the number of impulses in 200-ms intervals. (Teeter and Bennett 1981)

1992b). The potencies of L-GLU and its agonists, defined as the concentrations which gave 50% of maximal frequency increase, were of the order of 7 μM (Q), 25 μM (KA) and 0.6 mM (L-GLU). The excitatory effects of L-GLU persisted in receptors suppressed by high Mg^{2+}, indicating that L-GLU was acting at the postsynaptic site. Sensitivity to bath-applied L-GLU obtained in these experiments corresponds well to the behavior of other synapses of acousticolateral receptor organs. A similar sensitivity has been reported for the organ of Corti (Bobbin 1979; Comis and Leng 1979; Jenison et al. 1986), ampullary electroreceptors of the marine catfish Plotosus (Okano and Obara 1988), mechanoreceptors in the vesticular apparatus of the frog (Annoni et al. 1984) and lateral-line organ of Xenopus laevis (Bledsoe et al. 1983). The rank order in potency of the amino acid agonists in this synapse was similar to that found in the afferent synapse of electroreceptors in another teleost, the marine catfish Plotosus (Okano and Obara 1988). Both these preparations were similarly insensitive to applied NMDA. This argues against a role of the NMDA-type of postsynaptic receptors in synaptic transmission in ampullary electroreceptors of the catfish. The results are consistent with our current concept that Q AMPA- and KA-types, but not NMDA-type, membrane receptors may mediate the action of L-GLU in teleost electroreceptors.

A great contribution to the study of synaptic transmission in electroreceptors has been made by the investigation on marine catfish (*Plotosus*) receptors (Bennett and Obara 1986). L-GLU at a concentration of 10^{-4} to 10^{-9} M has been shown to produce an excitatory effect on these receptors (Higuchi et al. 1980; Okano and Obara 1988) (Fig. 69). Moreover, the observed L-GLU effects persisted in high-Mg^{2+} solution, indicating the existence of postsynaptic responses. The dose-response curves for L-GLU were identical in receptors where the resting activity was suppressed either by cathodal pulses or by high-Mg^{2+} solution (Okano and Obara 1988). The authors excluded any presynaptic L-GLU effect. Iontophoretic L-GLU application to the postsynaptic membrane of afferent fibers also evoked an excitatory effect with little desensitization (Nagai and Obara 1985). All excitatory amino acid agonists under study were classified into two groups in their ability to induce postsynaptic responses. The substances of the first group exerted an excitatory effect. These were Q, KA, L-GLU, and L-homocysteate, while L-ASP induced only a small slow long-latency response. The substances of the second group, including D-GLU, D-ASP, L-cysteate, L-cysteine sulfinate, and NMDA (10 mM), induced no responses (Okano and Obara 1988). Some aspartate analogs, L- and D-ASP, L-cysteate, and L-cysteinesulfinate, induced persistent potentiation of L-GLU responses. The antagonists of Q and NMDA types amino acid excitatory receptors, GDEE, DLAA, and APV, had no effect on either the resting activity or GLU-induced responses. It is noteworthy that KENYA slightly suppressed both resting and evoked activity (Okano and Obara 1988) and evoked dose-dependent inhibition of L-GLU- and Q-induced responses. L-GLU responses seem to be more resistant to KENYA than those of Q.

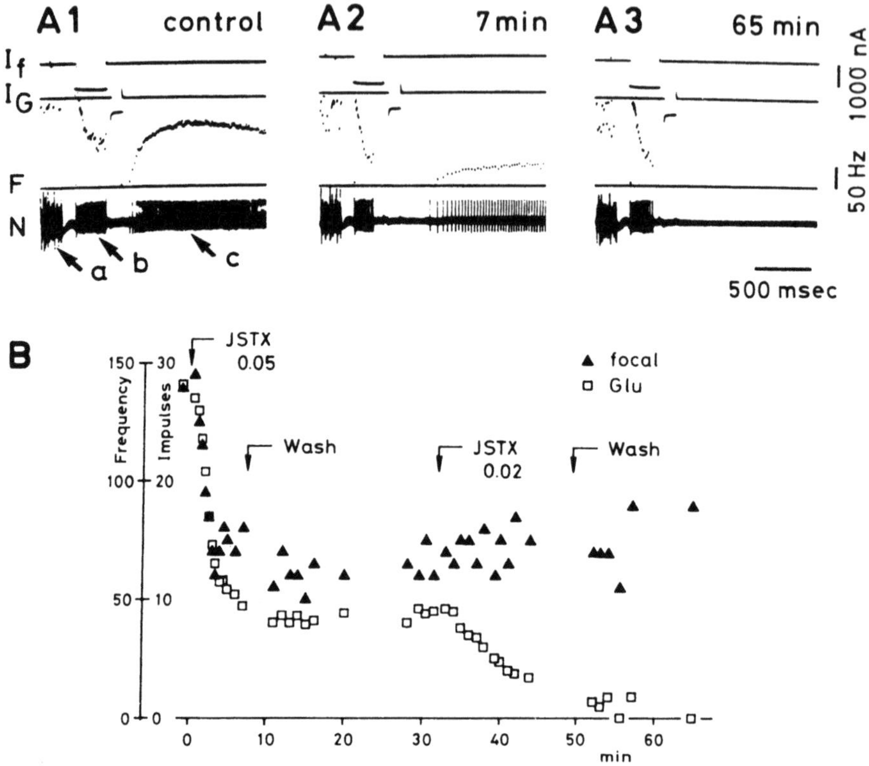

Fig. 69A,B. *Joro* spider toxin (JSTX) blocks the glutamate (GLU) response in ampullary electroreceptors of *Plotosus*, but not the resting discharges. **A** Sample of records with maximal stimuli for single units, before (**A1**) and after (**A2** and **A3**) bath application of JSTX. Single unit GLU response (*c*) induced iontophoretically (I_G, 100 ms) while multiunit resting discharges (*a*) were suppressed by hyperpolarizing the ampullae with a long current pulse. Focal response (*b*) induced by I_f (250 ms). Instantaneous frequency display (*F*). Note that resting discharges (*a*) and the focal responses (*b*) persist in JSTX, while the GLU response (*c*) was gradually and irreversibly blocked. **B** Effects of JSTX on response amplitudes plotted against elapsed time in minutes. *Ordinate*. response amplitudes in peak frequency for the GLU response (*squares*) and in impulse counts (250 ms) for the focal response (*triangles*). (Nagai et al. 1984. Reprinted with permission from Brain Research v. 300, pp. 185, 1984)

On the other hand, there are data contradicting L-GLU involvement in synaptic processes in *Plotosus* electroreceptors. For instance, *Joro* spider toxin (JSTX), blocking postsynaptic potentials and responses to L-GLU, KA, and Q application in lobster neuromuscular synapse (Kawai et al. 1982), did not influence the background and evoked *Plotosus* receptor activity but did block excitatory L-GLU effect (Fig. 69) (Nagai et al. 1984). The electrical stimulation of receptors led to endogenous neurotransmitter release into external solution capable of activating other ampullae, but biochemical

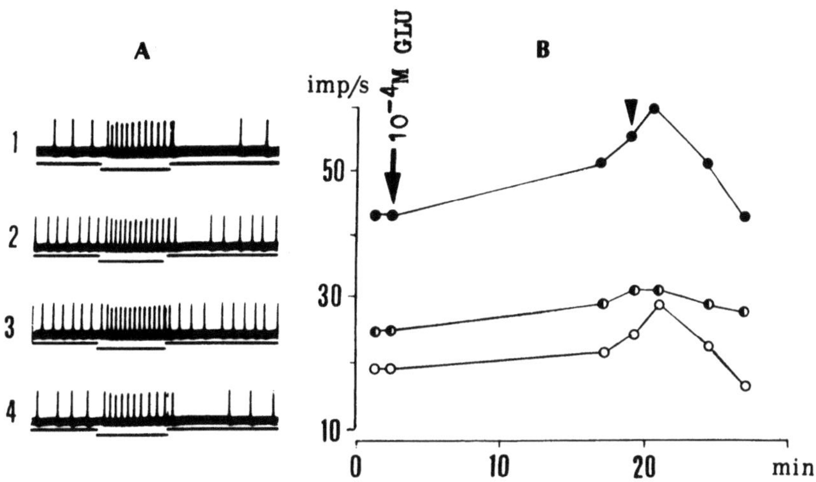

Fig. 70A,B. Effect of L-GLU on background and electrically evoked afferent activity of the ampullae of Lorenzini. **A** Upper beam: unit activity in afferent nerve. *1* response to electrical stimulation in normal solution; *2* and *3*, in solution containing 10^{-6} M L-GLU (after 3 and 5 min of drug application); *4*, in normal solution (after 15 min). *Lower beam* electric stimulus of 1.5 μA and 250 ms duration. **B** Changes of frequency in background (*black circles*) and electrically evoked activity (*white circles*) as well as reaction values (differences between evoked and background frequency (*black/white circles*). *Arrowheads* indicate application of normal solution

analysis failed to detect L-GLU in this medium (Umekita et al. 1980). The results of the study on neurotransmitters in the ampullary electroreceptors provide pharmacological evidence opposing the GLU hypothesis. According to Nagai and Obara (1985), the excitatory effect of L-GLU and its agonists is probably due to interaction with extrasynaptic receptors of unmyelinated nerve terminals. The above-mentioned results with JSTX may be explained by its ability to block only open channels (Jackson and Usherwood 1988). If this is the case, its blocking efficacy could depend on the number of open channels. Therefore, use-dependence experiments are necessary to clarify this problem.

In *Plotosus* receptors, strong excitatory pulses induced fast and slow PSP (Okano 1988). Slow PSP was suppressed by KENYA and γ-DGG and markedly prolonged by L-ASP. In contrast, these antagonists had little effect on fast PSP. In these electroreceptors Okano (1988) postulated two types of postsynaptic receptors corresponding to L-GLU receptors and to receptors with still unknown properties, respectively. On the basis of these data, a two-transmitter hypothesis has been advanced (Okano 1988).

L-GLU and L-ASP effects on the synaptic processes in skate electroreceptors (ampullae of Lorenzini) have been studied in detail (Steinbach 1974; Akoev et al. 1980b, 1981; 1984c, 1987b, 1991a, 1989; Akoev et al. 1991a). L-GLU and L-ASP in a concentration of 10^{-4} to

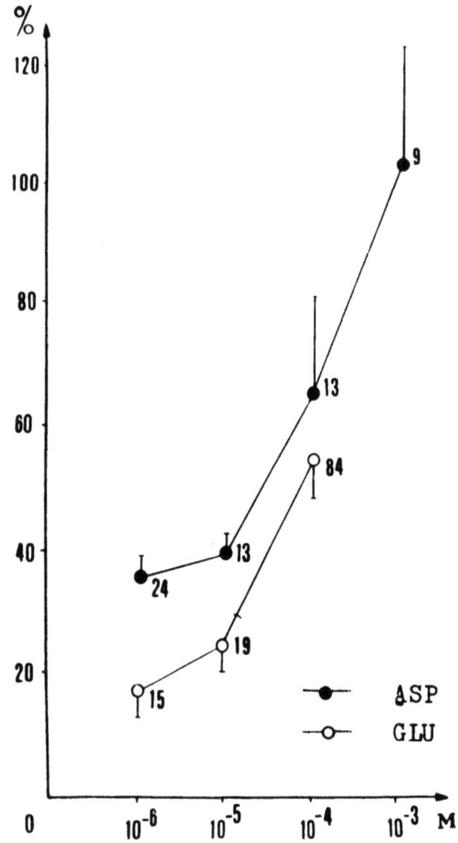

Fig. 71. Frequency response-concentration curves on application of EAAs to the afferent synapse of the ampullae of Lorenzini of the skate. *Figures* (at *points*) indicate number of tested afferents at a given concentration. *Ordinate* frequency values (in percent) after 3 min amino acid application. *Vertical bars* represent SEM

10^{-7} M induced a depolarization of the postsynaptic membrane and a reversible increase in the frequency of the background and evoked activity of the receptors in a dose-dependent manner (Figs. 70, 71, 72). L-ASP in most cases (81% of the receptors) had an excitatory effect, in 11% of the cases there were no changes, and in 8% the resting firing rate was decreased. The latter effect was observed in nerve fibers with a relatively high level of background activity. Similar data were obtained with L-GLU. An excitatory type of receptor was observed here also in the overwhelming majority of receptors to L-GLU application, and in a minor number of experiments (10%) the frequency of impulse activity showed a decrease. The inhibitory effect of L-GLU was also noticed in the *Plotosus* electroreceptors (Higuchi et al. 1980). The excitatory effect of amino acids was determined by their concentration; a rise in concentration produced an increase in the reaction

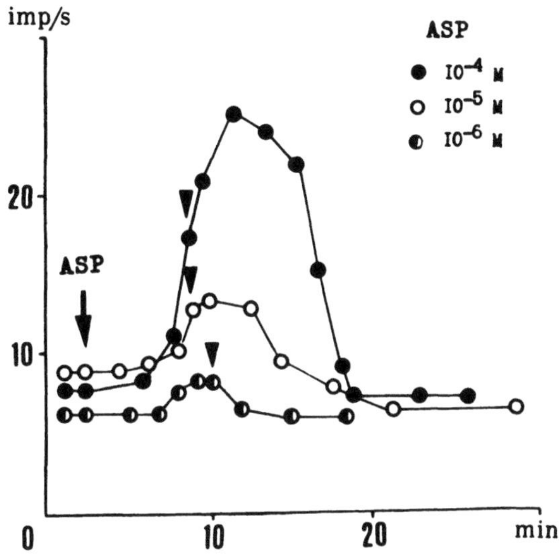

Fig. 72. Effect of L-aspartate (*ASP*, *arrow*) of different concentrations on background activity of the ampullae of Lorenzini. *Arrowheads* indicate application of normal solution

and a decrease in latency of stimulus-evoked responses. An initial increase in firing rate in the course of perfusion with amino acids was followed by a decrease in firing frequency.

The dose-response characteristics proved to be linear in concentration of 10^{-3} to 10^{-6}M (Fig. 71). The effect induced by L-ASP was somewhat higher than that of L-GLU. The threshold concentration for both amino acids in the ampullae of Lorenzini was 10^{-7}M, coinciding with those for other types of electroreceptors (Higuchi et al. 1980; Teeter and Bennett 1981), and is about three orders of magnitude lower than in other receptors of the acousticolateral system (Bledsoe et al. 1983; Bobbin et al. 1984). The question arises as to whether there is a special mechanism which may enhance electroreceptor sensitivity to amino acids. So far, this question has not been resolved. The high sensitivity of the electroreceptor synapse to a change in potential may be assumed to determine also its high sensitivity to amino acids since there is evidence on the existence of presynaptic glutamate receptors (Giorguieff et al. 1977). These receptors may act as autoreceptors, controlling the release of the neurotransmitter by a positive feed-back.

To determine the site of amino acid action on synaptic structures of the ampullae of Lorenzini, experiments with combined action of a solution with high Mg^{2+} and amino acids were performed. The results of one such experiment are depicted in Fig. 73. L-ASP is shown to cause a marked rise in the frequency of impulse activity restored in normal solution. A rise in Mg^{2+} concentration led to a complete reduction of the firing rate in nerve fibers. In this period receptors responded to neither an electrical nor a temperature

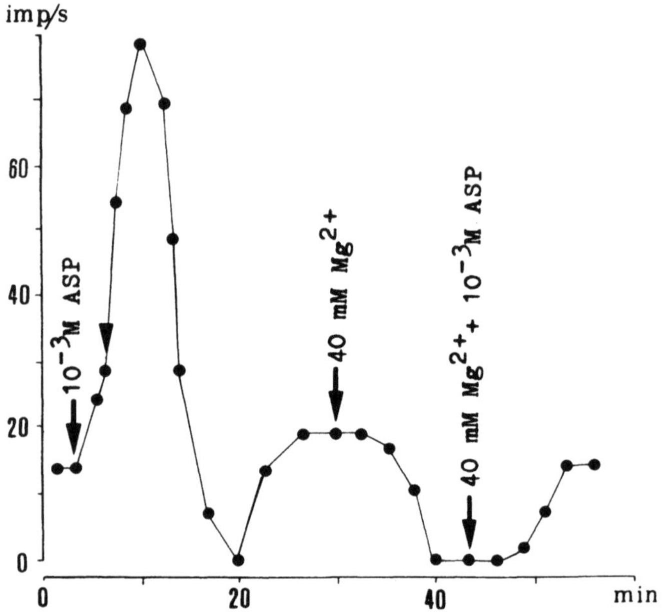

Fig. 73. Effect of 10^{-3} M L-ASP on background activity of ampullae of Lorenzini in high (40 mM) Mg^{2+} solution. *Arrowheads* indicate application of normal solution

stimulation. Yet an addition of L-ASP to the high Mg^{2+} solution induced firing in nerve fibers, and a sustained level of background activity was set, determined by amino acid concentration. It is to be noted that under these conditions, the sensitivity to electrical stimuli was abolished, and the receptor responded to a temperature stimulation with an inverted impulse reaction having a positive temperature coefficient. Such responses differed drastically from the responses in normal solution (Akoev 1990). The ability of the postsynaptic membrane to maintain impulse activity for a long period in L-GLU solution testifies to the total absence of desensitization processes in the postsynaptic receptors in the ampullae of Lorenzini. This property may be important for tonic receptors to maintain high receptor sensitivity in continuous transmitter concentration.

Besides the postsynaptic effect, the EAA seem also to be capable of presynaptic action (Valli et al. 1985; Prigioni et al. 1990). Indeed, during external perfusion, L-GLU and L-ASP induced a dynamic rise in firing rate, with a subsequent decrease in activity. However, their time course and the degree of their reactions differ substantially from those of the pure postsynaptic effects observed in Mg^{2+}-rich solution; when slow tonic changes in the firing rate took place. Moreover, following L-GLU and L-ASP application to the normal solution especially in high doses, the recovery of activity in normal solution is reduced, and the receptor excitability may be irreversibly lost. Similar effects of EAA in ampullary electroreceptors were also reported (Higuchi et al. 1980; Nagai and Obara 1985; Teeter and Bennett 1981). Thus, the receptor reaction to bath-applied amino acids differs from that evoked by natural stimulation of receptors. At present, however, the mechanisms of bath-applied EAA effects are unknown, although the presynaptic membrane and supporting cells may be involved. On the contrary, in *Plotosus* electroreceptors a presynaptic effect of L-GLU and agonists was excluded (Okano and Obara 1988).

The sensory synapse of the ampullae of Lorenzini appeared to be sensitive to EAA agonists. This synapse has been found to be most sensitive to KA (Akoev et al. 1986a, 1987a, 1989, 1991a). The threshold concentration of this substance was 10^{-9} M. KA in the concentration range of 10^{-5} to 10^{-9} M increased the rate of spontaneous discharge, dose-dependently and reversibly, in units with low initial resting firing rate, while in units with a high level of resting activity KA caused a reduction in activity. Higher KA doses, however, induced a transient excitation followed by suppression. To determine the site of KA action on synaptic structures of the ampullae of Lorenzini, experiments were performed with a combined high Mg^{2+} concentration and KA action. A rise in Mg^{2+} concentration up to 20–30 mM led to a complete blocking of synaptic transmission and a loss of sensitivity to temperature and electrical stimuli (Fig. 74). Addition of KA to hypermagnesium solution caused a restoration of background activity, depending on its concentration. An essential difference in the KA effect from that of L-GLU was the fact that, in Mg^{2+}-rich solution, the former restored not only background activity but also the evoked activity (Fig. 74)

The mechanism of KA influence on nerve cells has long been discussed in the literature (Nadler 1979; Coyle 1983; Bledsoe et al. 1981a). At least three independent KA effects were reported: first, a weak postsynaptic excitatory effect or a potentiation of the L-GLU effect; second, a presynaptic effect related to the activation of L-GLU secretion; and finally, an inhibitory effect on the GLU uptake system (Foster and Fagg 1984; Coyle 1983). As to the presynaptic effect, KA may interact with GLU receptors located on the presynaptic membrane of the ampullae of Lorenzini and may selectively cause release of a natural transmitter. Such a mechanism has been demonstrated on the synaptosomes of the rat brain (Pastuszko et al. 1984). KA was shown to interact with the GLU receptors of the presynaptic membrane and, via voltage-dependent processes, to raise its permeability for Ca^{2+}, thereby activating a transmitter release. Furthermore, KA stimulates only GLU and ASP release, in contrast to potassium ions, which also activate a release of GABA and other amino acids.

On the other hand, a possible postsynaptic KA effect on the synapse of the ampullae of Lorenzini can not be denied, as has been proved for the hair cell synapse of the guinea-pig cochlea (Bledsoe et al. 1981a). According to Johnston et al. (1979), KA, possessing a "rigid" molecular structure, can interact with a subclass of glutamate receptors highly selective for a partly

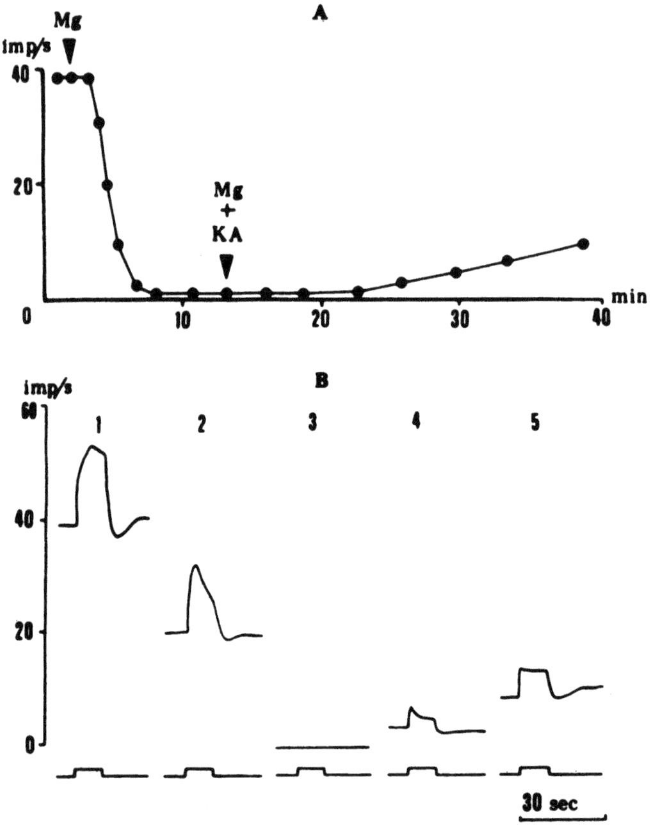

Fig. 74A,B. Effects of 10^{-6} M KA on the resting (**A**) and evoked (**B**) firing rate of the ampullae of Lorenzini in Mg^{2+}-rich solution. **B** *1* response to electric pulse in normal solution; *2,3* responses in normal solution containing 50 mM Mg^{2+} (3 and 10 min, respectively); *4,5* responses in Mg^{2+}-rich solution plus 10^{-6} M KA (25 and 35 min, respectively). *Upper beam* firing frequency, imp/s; *lower beam* stimulus current. Intensity of applied current was 0.15 μA and 0.5 μA in high Mg^{2+} solution. Duration of electrical stimuli 10 s. (Akoev et al. 1987a) (Reprinted with permission from Comp. Biochem. Physiol. v. 88A, Akoev et al. Effect of kainic acid on synaptic transmission in electroreceptors (the ampullae of Lorenzini) of skates. Copyright 1987, Pergamon Press PLC)

extended form of L-GLU. These KA-receptors constitute only about 10% of the total number of L-GLU binding receptors.

Addition of Q and NMDA to a normal solution elicited a transient and dose-dependent increase in the firing rate of the ampullae of Lorenzini (Figs. 75, 76). For both agonists, the responses to the external perfusion produced a rise in the firing rate which in the course of continuous perfusion experienced a gradual decrease. High doses of these agonists induced a transient excitation followed by suppression. This was probably due to the depolarizing inactivation of the spike generation mechanism at the afferent

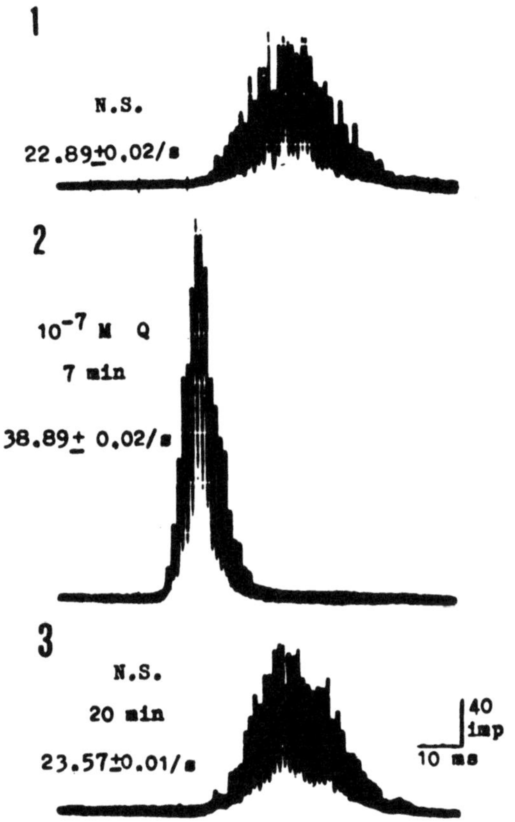

Fig. 75,1–3. Interval spike histograms of background activity in afferent ampullary fibers before (**1**), at the time (**2**), and after (**3**) application of 10^{-7} M quisqualic acid (*Q*)

nerve terminals. The threshold concentration for Q and NMDA were 10^{-8} M and 10^{-5} M, respectively. Thus, the sensory synapse of the ampullae of Lorenzini appeared to be sensitive to excitatory amino acid agonists. Thus, the order of potencies for agonists was KA > Q > L-GLU (=ASP) > NMDA. A similar rank order in potencies of the amino acid agonists was observed in the primary afferents of the vestibular system of the axolotl (Soto and Vega 1988). *Plotosus* receptors were shown to be sensitive to Q rather than to KA (Okano and Obara 1988).

Q- and NMDA-evoked responses were shown to be sensitive to Mg^{2+}. In high Mg^{2+} solution the excitatory responses to NMDA were depressed completely, while responses to Q were reduced to a lesser extent (Fig. 77). It is noteworthy that under both the complete blockade of synaptic transmission and a slight decrease in resting firing rate in Mg^{2+}-high solution, NMDA application was quite inefffective. In contrast, Q-evoked responses proved to be relatively resistant to Mg^{2+} action. In high-concentration Mg^{2+}

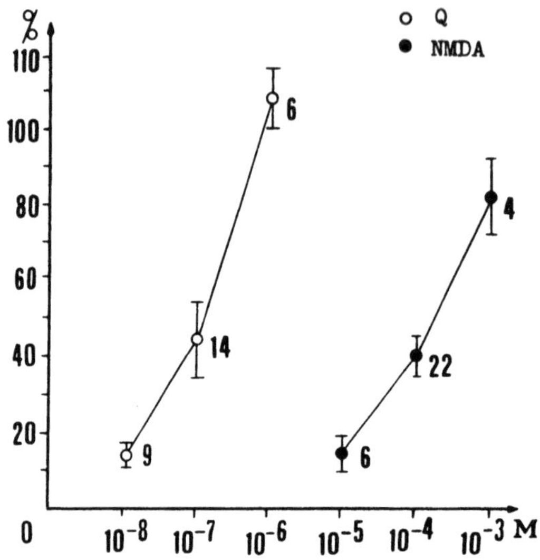

Fig. 76. Frequency response-concentration curves on application of Q and NMDA to the afferent synapse of the ampullae of Lorenzini. *Figures at points* indicate number of afferents tested at a given concentration. *Vertical bars* represent SEM. *Ordinate*. frequency values (in percent) after 3 min application of agonists

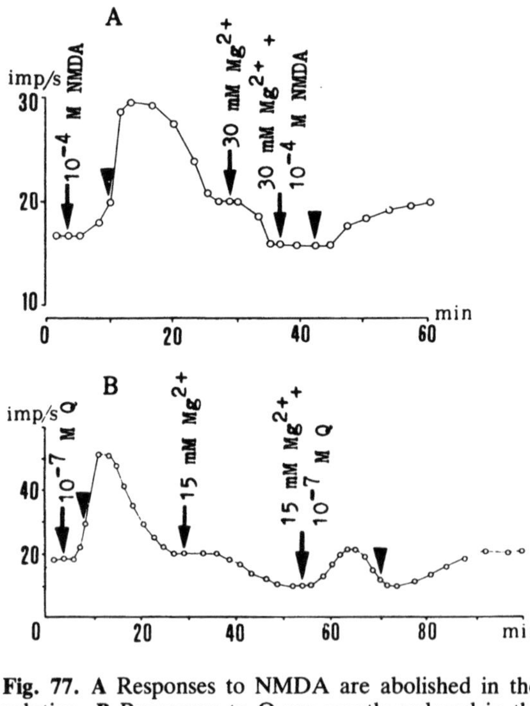

Fig. 77. A Responses to NMDA are abolished in the presence of high (30 mM) Mg^{2+} solution. **B** Responses to Q are greatly reduced in the presence of high (15 mM) Mg^{2+} concentration. *Arrowheads* indicate application of normal solution

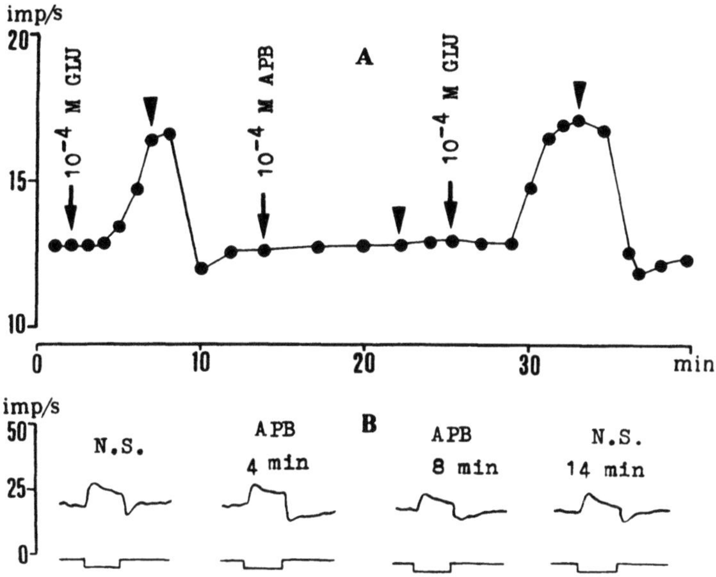

Fig. 78A,B. Lack of effect of 10^{-4} M 2-amino-4-phosphonobutyric acid (*APB*) on background, L-GLU-induced (**A**) and electrically evoked (**B**) activity of the ampullae of Lorenzini. *Lower beam* electrical stimulus, 1.5 μA, 10 s. *Arrowheads* indicate application of normal solution

solution, Q evoked firing in afferent fibers under complete or partial suppression of resting activity.

There is ample evidence that L-GLU and L-ASP are mixed agonists producing their effects at NMDA and non-NMDA postsynaptic receptors (Watkins and Evans 1981). The NMDA receptor is selectively antagonized by DLAA, APV, APB, and the ion channel is selectively blocked by extracellular Mg^{2+} ions and MK-801. In the ampullae of Lorenzini, DAA and APB did not affect either resting or evoked activity (Fig. 78). The existence of NMDA postsynaptic receptors is proved by the depressing effect of APV and Mg^{2+} on resting and electrically evoked activity as well as by the total blockade of NMDA-evoked responses; on the contrary, APV had less effect on L-ASP-induced responses (Figs. 79, 80) (Akoev et al. 1989, 1991b). The relative insensitivity of the L-ASP responses to NMDA receptor antagonists suggests that L-ASP acts at receptors distinct from those reacting to NMDA, i.e., Q and KA.

Selective antagonosts distinguishing the KA and Q effects have not been found so far, with the exception of GDEE, which predominantly affects Q-evoked responses (Watkins 1981). At a concentration of 10^{-4} to 10^{-6} M, GDEE and GDME reduced the frequency of resting and electrically evoked activity in the afferent fibers of the ampullae of Lorenzini (Fig. 81) (Akoev et al. 1981, 1987c, 1988a, 1991b). In the range 10^{-4} to 10^{-6} M, the degree of

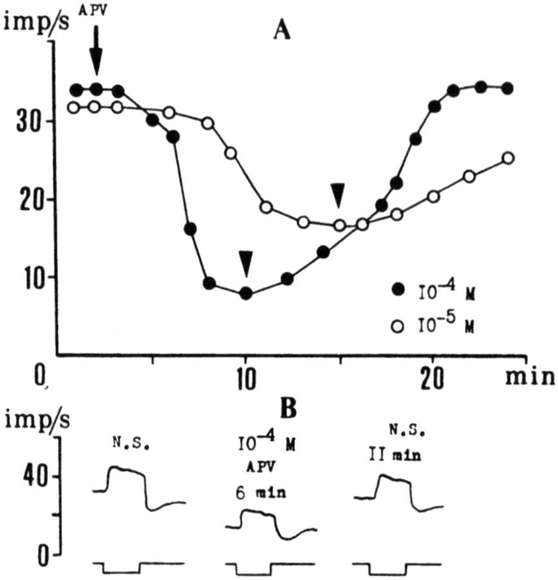

Fig. 79A,B. Effect of 2-amino-5-phosphonovaleric (*APV*) on resting (**A**) and electrically evoked (**B**) activity of the ampullae of Lorenzini. *Arrowheads* indicate application of normal solution

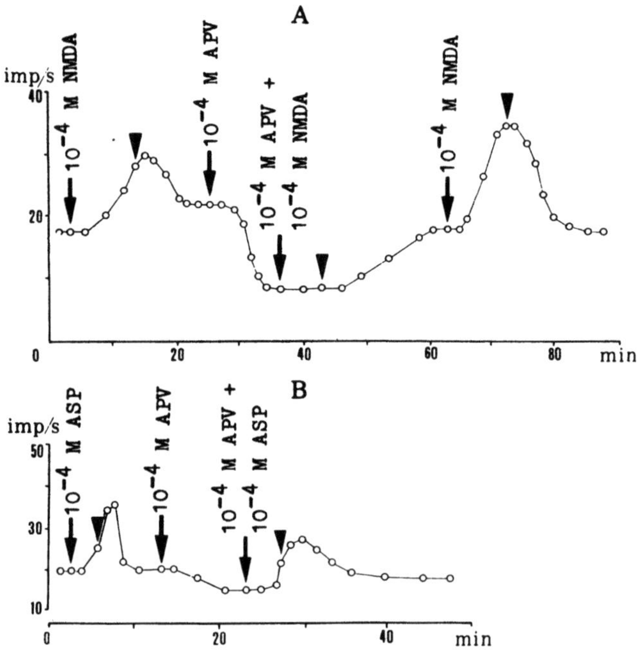

Fig. 80A,B. Effect of APV on the responses evoked by NMDA (**A**) and L-ASP (**B**) in the ampullae of Lorenzini. APV abolished NMDA response but had no effect on L-ASP response. *Arrowheads* indicate application of normal solution

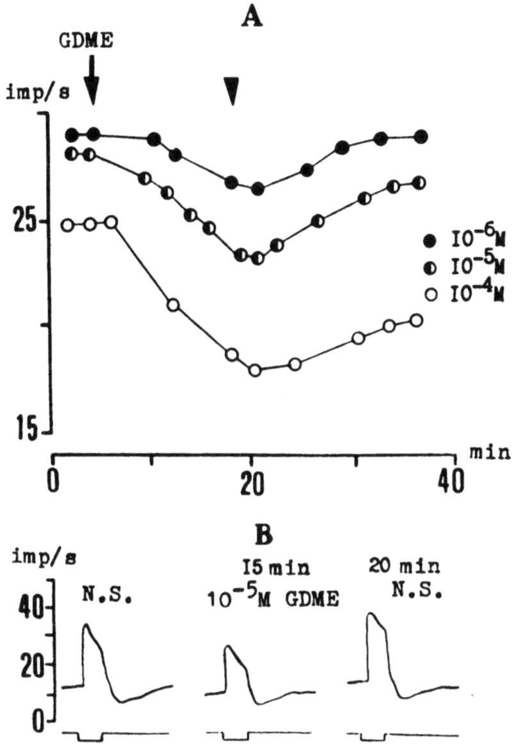

Fig. 81A,B. Effect of L-glutamic acid dimethylester (GDME) of different concentrations on the background (**A**) and electrically evoked (**B**) activity in the afferent nerve of ampullae of Lorenzini. *Arrowheads* indicate application of normal solution

reduction was concentration-dependent. At a concentration of 10^{-3} M, both antagonists usually evoked a complete and often irreversible cessation of impulse activity. The influence of these antagonists supports the conclusion regarding the postsynaptic effects of amino acids. Both chemicals antagonized the postsynaptic amino acid-induced depolarization. For example, the responses to L-GLU were reduced by 50% in a 10^{-5} M GDME solution (Fig. 82). It should be noted that GDEE acted directly upon the postsynaptic receptors, since in high-concentration Mg^{2+} solution, GDEE reversibly blocked postsynaptic responses evoked by EAA. In these experiments, the background activity was first blocked by the high Mg^{2+} solution, and after being restored by L-GLU, it was blocked again by the addition of GDEE (Fig. 83). Thus, non-NMDA Q-type receptors may also exist in the afferent synapses of the ampullae of Lorenzini, since Q evoked strong responses relatively resistant to high Mg^{2+} solution, and effectively blocked by GDEE.

KENYA, PDA, and γ-DGG are broad-spectrum antagonists affecting the responses provoked by KA and Q (Foster and Fagg 1984; Perkins and Stone

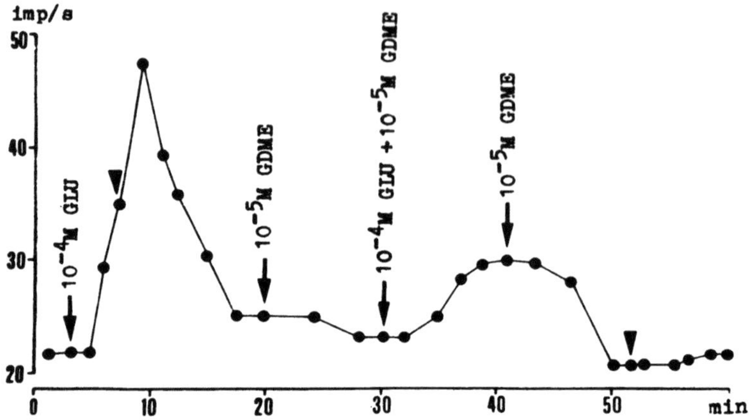

Fig. 82. Effect of 10^{-5} M GDME on response to L-GLU. The response is reduced by 50% in the presence of GDME. *Arrowheads* indicate application of normal solution

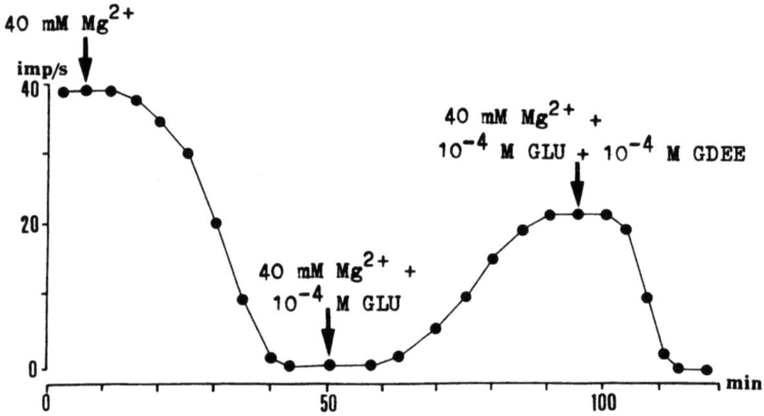

Fig. 83. Effect of 10^{-4} M L-glutamic acid diethylester (*GDEE*) on postsynaptic responses to L-GLU application in the ampullae of Lorenzini where background activity was first suppressed by high Mg^{2+} concentration. L-GLU response was abolished by addition of 10^{-4} M GDEE

1982; Watkins and Evans 1981; Fonnum 1984). In the ampullae of Lorenzini, PDA at a concentration of 10^{-4} to 10^{-6} M affects both resting and amino acid-induced responses (Akoev et al. 1988a, 1991b). PDA reduced L-GLU-induced responses to a lesser degree than those induced by L-ASP (Fig. 84). On the contrary, KENYA had a stronger effect on the L-GLU-induced response than on L-ASP-induced (Fig. 85). DGG was found to be an

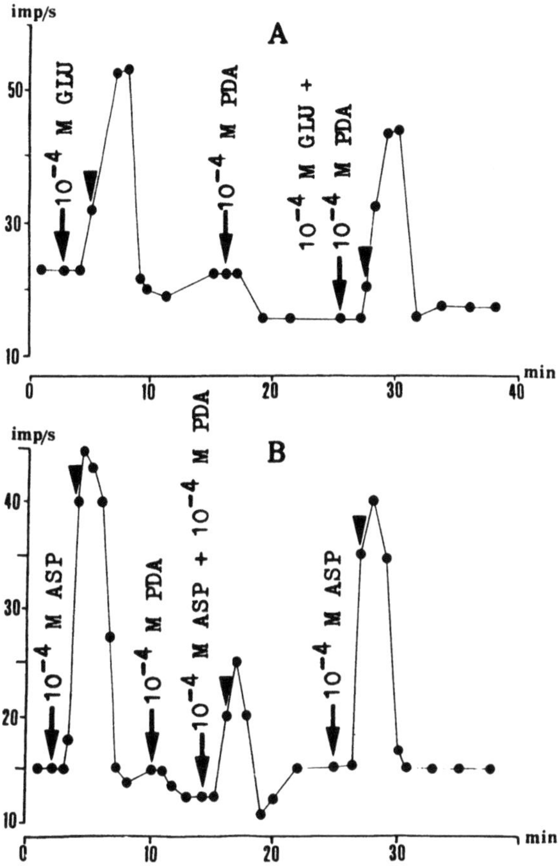

Fig. 84A,B. Effect of *cis*-2,3-piperidinedicarboxylate (*PDA*) on L-GLU (**A**) and L-ASP (**B**) responses. Note that the L-GLU response is little affected by PDA, while that of L-ASP is very much affected. *Arrowheads* indicate application of normal solution

effective antagonist of synaptic transmission in the ampullae of Lorenzini. At a concentration of 10^{-4} to 10^{-6} M, it greatly reduced the resting activity in the afferent fibers (Fig. 86).

L-serine-O-phosphate (SOP) in the ampullae of Lorenzini was shown to depress both resting and evoked activity at a threshold concentration of 10^{-7} M. SOP reduced postsynaptic responses evoked by L-GLU and NMDA, while those induced by Q and KA were unaffected (Akoev et al. 1992).

Recent studies have demonstrated with [^{3}H] GLU-binding activity of plasma membranes of the ampullae of Lorenzini that these membranes exhibit predominantly one class of binding sites (Akoev et al. 1986a). The dissociation constant (K_d) was shown to be 0.2 μM, the corresponding binding site densities (B_{max}) being 30 pmol/mg protein. The K_d and B_{max} values

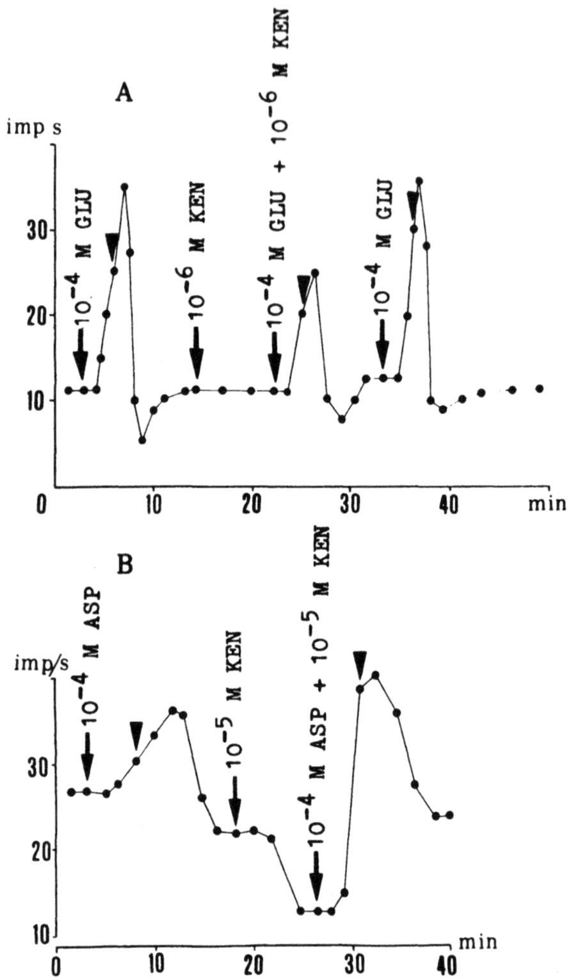

Fig. 85A,B. Effect of kynurenic acid (*KEN*) on L-GLU (**A**) and L-ASP (**B**) responses. At 10^{-6}M concentration, KEN depressed response to L-GLU; it did not affect L-ASP response even at a concentration of 10^{-5}M. *Arrowheads* indicate application of normal solution

determined in these experiments are in good agreement with those of plasma membranes of brain tissue (Foster and Fagg 1984).

A detailed study was carried out to elucidate the role of putative neurotransmitters in synaptic transmission in electroreceptors (Akoev and Andrianov 1989). A possible role of GABA, adrenaline, noradrenaline, ACh, dopamine, taurine, serotonin, and histamine was examined.

There are a number of reports concerned with the effect of ACh, on synaptic transmission in electroreceptors. ACh and d-tubocurarine were

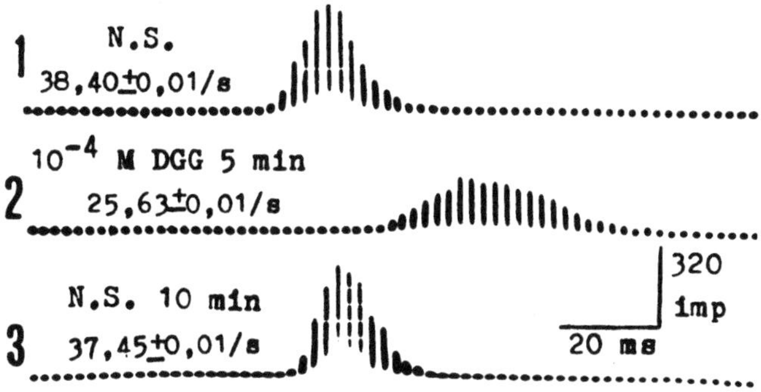

Fig. 86,1–3. Effect of γ-D-glutamylglycine (*DGG*) demonstrated on interspike histogram of an afferent fiber of the ampullae of Lorenzini obtained in normal solution (**1**,**3**) and in the presence of DGG (2)

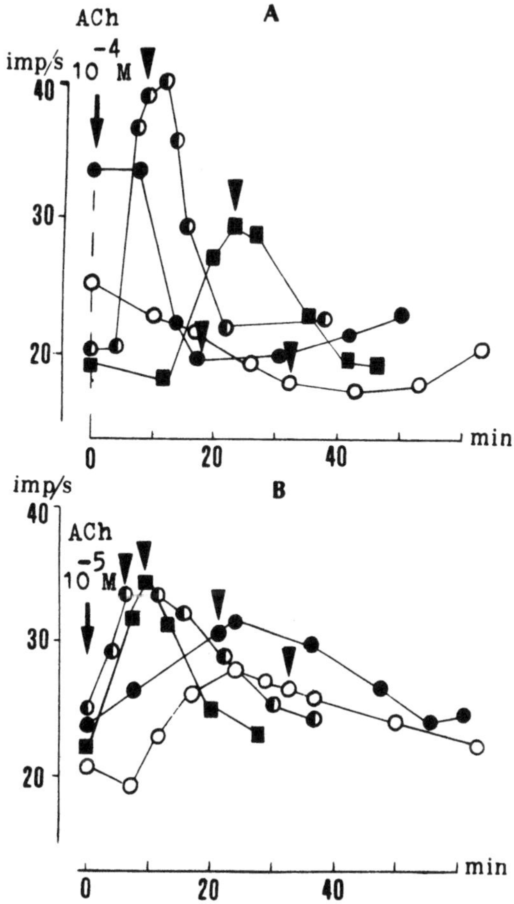

Fig. 87A,B. Frequency changes in beckground activity of the ampullae of Lorenzini provoked by solutions containing 10^{-4} M (**A**) and 10^{-5} M (**B**) acetylcholine (*ACh*). Data for four (*different symbols*) ampullae of Lorenzini. *Arrowheads* indicate application of normal solution shown to produce only negligible changes in the spontaneous and evoked discharges of tonic electroreceptors in the transparent catfish, *Kryptopterus* (Teeter and Bennett 1981), and in the marine catfish, *Plotosus* (Bennett and Obara 1986). The effects of ACh and cholinolytics on synaptic processes in the ampullae of Lorenzini of skates have also been investigated (Akoev et al. 1982, 1983). At a concentration of 10^{-5} M, ACh exerted a strong excitatory effect on these receptors (Fig. 87), but with higher concentration (10^{-3}–10^{-4} M) the effect of ACh depended on the value of the resting activity. If resting discharge was low (less than 22–25 imp/s), ACh caused an increase in the frequency of resting activity; in the case of the initial frequency being high (more than 25 imp/s), ACh reduced the resting discharge. Such effects of ACh may be attributed in principle to its influence on pre- and postsynaptic membranes. However, by blocking synaptic transmission with high-concentration Mg^{2+} solution, ACh caused no impulse activity in afferent fibers (Fig. 88). This seems to indicate that ACh has no effect on the permeability of ionic channels of the postsynaptic membrane. Therefore, the observed ACh effects on the resting activity of the ampullae of Lorenzini may be explained by its presynaptic action. Indeed, an addition of ACh into the solution may depolarize the presynaptic membrane, and thus may raise the probability of transmitter release, as in the case of other hair cell synapses (Bobbin et al. 1985; Norris et al. 1988; Bernard et al. 1985). The cholinergic antagonists, d-tubocurarine and atropine, did not affect synaptic transmission in the ampullae of Lorenzini. Thus, ACh is unlikely to be an afferent transmitter in electroreceptors.

Currently, a considerable amount of evidence suggests that GABA is not used as a neurotransmitter by electroreceptors. GABA did not change the

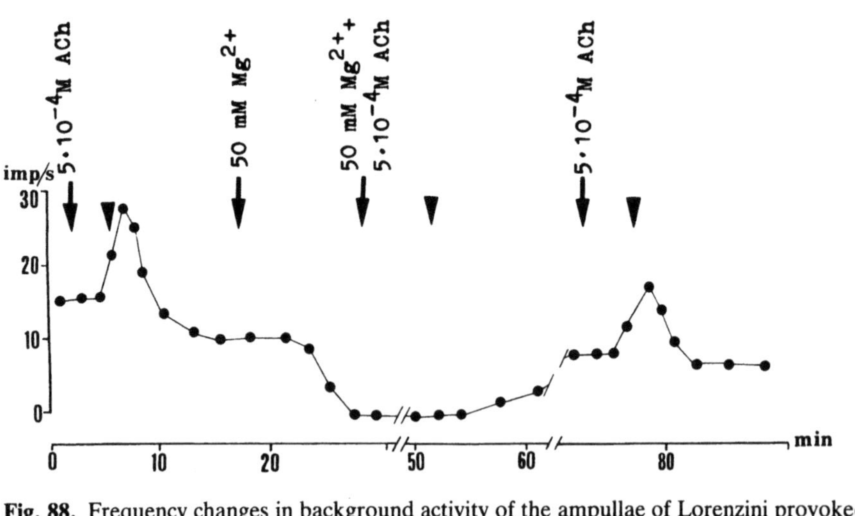

Fig. 88. Frequency changes in background activity of the ampullae of Lorenzini provoked by solutions containing 10^{-4} M ACh and 50 mM Mg^{2+}. *Arrowheads* indicate application of normal solution

background activity of the *Plotosus* electroreceptors (Teeter and Bennett 1981). On the other hand, this agonist, at a concentration of 10^{-4} to 10^{-5} M, reduced the resting firing rate and did not affect the evoked activity of the ampullae of Lorenzini of the skate (Fig. 89A) (Akoev et al. 1985). GABA did not change the EPSP amplitude in these electroreceptors (Steinbach 1974). Picrotoxin, being a GABA receptor antagonist, at a concentration of 10^{-4} M, slightly reduced the background activity in afferent fibers, producing no effect on evoked activity (Fig. 89B).

The possible role of catecholamines as afferent neurotransmitters in electroreceptors has been studied in several works. Studies on the ampullae of Lorenzini in skates demonstrated that catecholamines reduced the frequency of the resting activity of the receptors (Fig. 90) (Akoev et al. 1985). Adrenoreceptor antagonists (phentolamine and propranolol) led to some decrease in the frequency of background activity and did not block receptor responses to electrical stimuli. It is noteworthy that β-adrenoblocker, propranolol (10^{-5} M), did not influence the inhibitory effect of adrenaline (Fig. 91A). On the other hand, an α-adrenoblocker, phentolamine, completely blocked the effect of adrenaline on the ampullae of Lorenzini (Fig. 91B). No obvious effects were produced by catecholamines on electroreceptors of the marine catfish *Plotosus* (Bennett and Obara 1986). Thus, it

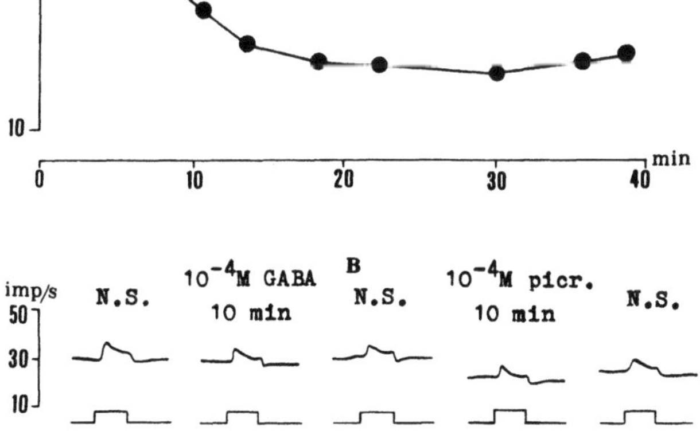

Fig. 89A,B. Effect of GABA and picrotoxin on background (**A**) and evoked (**B**) activity of ampullae of Lorenzini. **A** Effect of GABA (10^{-4} M) for two (*circles* and *dots*) receptors. **B** *Upper beam*, responses to electrical stimulation; *lower beam* cathodal stimulus of 10 s duration and 0.5 μA intensity. *Arrowheads* indicate application of normal solution

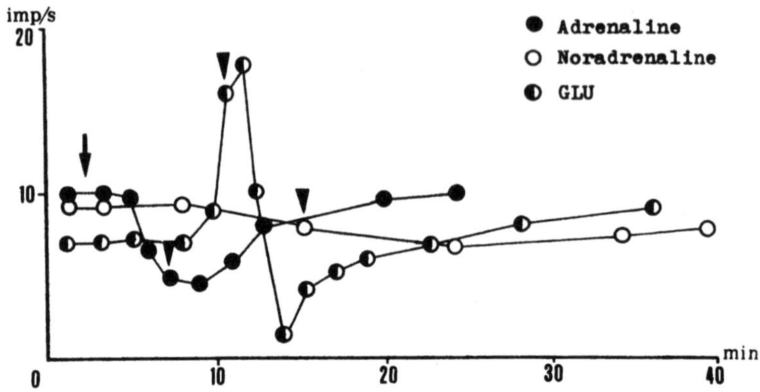

Fig. 90. Effect of 10^{-4}M adrenaline, 5 × 10^{-5}M noradrenalin, and 10-yM L-GLU (*arrow*) on background activity of the ampullae of Lorenzini. *Arrowheads* indicate application of normal solution

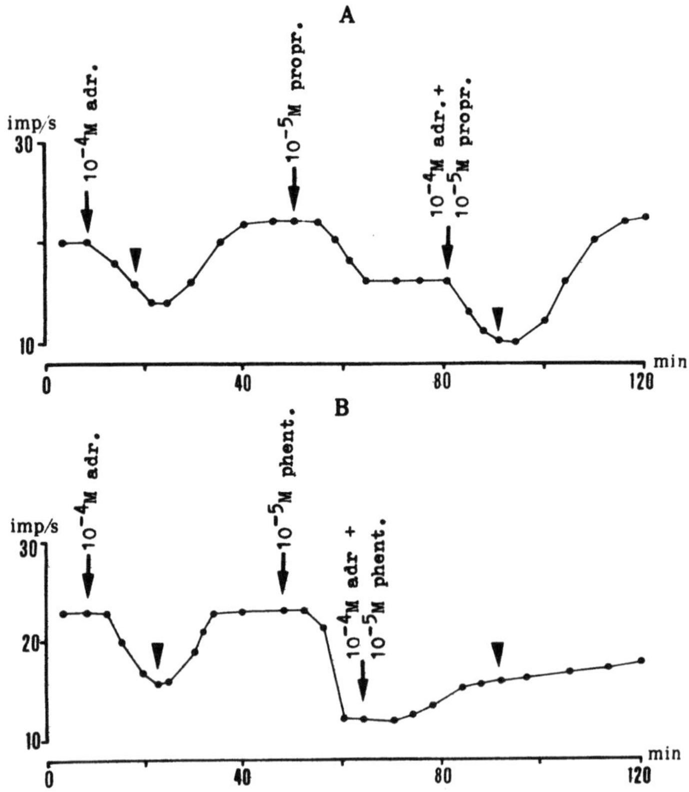

Fig. 91A,B. Frequency changes in background activity of ampullae of Lorenzini provoked by solutions containing 10^{-4}M adrenaline (*adr*) and 10^{-5}M propronalol (*propr*) (**A**) and 10^{-4}M adrenaline (*adr*) and 10^{-5}M phentolamin (*phent*) (**B**). *Arrowheads* indicate application of normal solution is probable that catecholamines may not be considered as putative neurotransmitters in electroreceptors.

According to available data, dopamine, serotonine, taurine, and 5-hydroxytriptamine cannot be considered as putative transmitters in the sensory synapse of the ampullae of Lorenzini of skates. Taurine was shown to exert an inhibitory effect only on the ampullae of Lorenzine (Okano et al. 1984; Akoev et al. 1985). Serotonine and dopamine at a concentration of 10^{-3} to 10^{-5}M, as well as their antagonists, i.e., metergoline and fluphenasine, did not influence the background and evoked activity in skate electroreceptors.

Thus, on the basis of the available data, we may conclude that L-GLU, L-ASP, or a structurally related substance are the most probable neurotransmitters in electroreceptors.

During the past two decades, considerable progress has been made toward understanding the mechanisms of synaptic transmission in receptors of the acousticolateral system. Throughout this system, the sensory receptors are the hair cells. Their fundamental morphology is similar to a marked degree from system to system and from species to species. A key function of the hair cells is to release a chemical neurotransmitter under adequate stimulation. Analysis of available experimental data indicates that ACh, serotonine, dopamine, norepinephrine, GABA, or taurine are not mediators in hair cells. To date, the best supported evidence shows that EAA (GLU and ASP), or a structurally related compound, are the most probable neurotransmitters. One may therefore suppose that there is only one neurotransmitter in different receptors of the acousticolateral system. On the other hand, there is a possibility that the neurotransmitter is not the same in this system. In this context, an alternative hypothesis would indicate that more than one transmitter (cotransmitters) is (are) released by given hair cells. In the ampullary electroreceptors of *Plotosus*, Okano (1988) recently postulated a two-transmitter hypothesis. It is well established that during stimulation or in high-concentration K^+ solution several chemicals are released from hair cells, supporting cells, or nerve endings. These compounds may act on receptors located on the hair cells, modulating the process of neurotransmitter release and thus they can alter the properties of postsynaptic receptors or unmask additional receptors and thereby the responsiveness of afferent fibers (Bledsoe et al. 1988). It is of interest that opiate peptides may play a modulatory role in hair cells. This point of view is supported by the observation that opiate peptides are known to suppress responses of brain neurons (Zieglgansberger 1980) and ampullae of Lorenzini to L-GLU (Andrianov et al. 1993) and to reduce GLU binding by synaptic membranes (Kuznetsov and Godukhin 1985). A modulatory role has also been postulated for adenosine and ASP (Bryant et al. 1987).

Although the available data support the hypothesis that L-GLU or L-ASP are the most probable neurotransmitters in hair cells of the acousticolateral system, many aspects of this problem, however, remain to be studied. The idea of these amino acids being neurotransmitters implies invariably the existence of specific postsynaptic receptors, although it is not quite clear why L-GLU and L-ASP produced effects at a rather high concentration in some synapses. At high concentrations, L-GLU can possibly affect the nonglutamergic postsynaptic receptors.

It should be mentioned that bath-applied L-GLU may influence both hair cells and afferent nerve fibers. L-GLU and L-ASP effects may be associated with the inhibition of the uptake of the natural neurotransmitter or its precursors. In addition, L-GLU is well known as exerting an excitatory effect on various neurons of the central and peripheral nervous systems, and in many cases its effects are not primarily associated with synaptic processes. L-GLU can also influence the presynaptic receptors by facilitating a release of the endogenous transmitter. This seems quite reasonable, since presynaptic GLU receptors are known to exist (Giorguieff et al. 1977) as well as GLU uptake by hair cells of organ of Corti (Gulley et al. 1979; Eybalin and Pujol 1983). It should also be pointed out that, although GLU and ASP have been shown to be present in the organ of Corti at high concentrations (Godfrey et al. 1976a), their mere presence is not sufficient for a neurotransmitter role. These amino acids are involved in a number of metabolic processes. In experiments on frog semicercular canal it has been shown that when the preparation was insensitive to subsequent injection of L-GLU it yet remains responsive to the endogenous transmitter released by the hair cells. (Guth et al. 1988) Additional experiments are still necessary to provide more information on the GLU hypothesis.

Pharmacological data have revealed the presence of receptors for EAA in hair cells of the acousticolateral system. Comparison of the biochemical and pharmacological data on the properties of postsynaptic receptors in various hair cells of the acousticolateral system shows significant differences between them, which are related to the existence of a receptor for NMDA in various

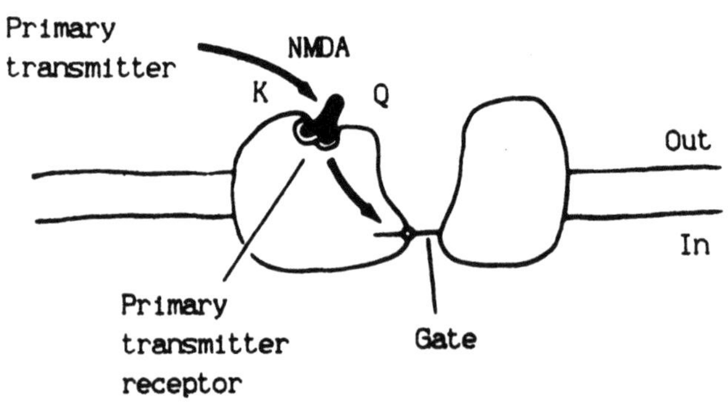

Fig. 92. Model of single-receptor/channel complex with separate ligand-binding sites for N-methyl-D-aspartate (*NMDA*), kainic acid (*K*), and quisqualic acid (*Q*) in the postsynaptic membrane of afferent fibers of the acousticolateral system

hair cells. Thus, in lateral-line NMDA, DLAA and APV were effective; in the semicircular canal of the frog and ampullary receptors of the skate both DLAA and APV depressed the effects of L-GLU as well as the background, and evoked activity. In contrast, in ampullary receptors of the marine catfish *Plotosus* and at the sensory synapses of hair cells of the mammalian organ of Corti, NMDA and selective antagonists of NMDA receptors, APV and DLAA, were ineffective. Thus, the receptors for KA and Q but not NMDA are present in hair cells of mammalian cochlea. As to the existence of Q and KA types of postsynaptic receptors in hair cells of the acousticolateral system, the available data are rather contradictory, especially concerning their sensitivity to various antagonists. Thus, these data indicate that there is little evidence to assume the existence of a single type of excitatory amino acid postsynaptic receptor among acousticolateral receptors.

On the contrary, in this organ at least two or three subtypes of postsynaptic receptors are present. What is the role of these multiple types of receptors? For example, NMDA receptors may be utilized during resting activity and at low levels of natural stimulation (Bledsoe et al. 1988). Conformation changes of postsynaptic receptors at low intensity stimulation could increase the affinity of Q and KA receptors to the neurotransmitter. Thus, receptors for Q and KA are activated by high intensity stimuli.

Recent evidence suggests that several different types of amino acid receptors exist in the neurons of the central nervous system, with numerous synaptic inputs. With respect to the acousticolateral receptors, morphological findings indicate that there is only one type of afferent synapse. Hence, there is no reason to believe in the existence of several separate types of postsynaptic receptors. If this is the case, we may assume that agonists act via a single-receptor/channel complex with multiple separate but not identical ligand-binding sites associated with different conductance levels (Fig. 92). Additional experiments are necessary to clarify this problem.

Conclusion

Hair cells are the receptor cells of the acousticolateral system, which is of great significance among the sensory organs in vertebrates. These cells are unique in terms of their high sensitivity to external stimuli, on the one hand, and their neurosecretory role and capability to synthesize and release a neurotransmitter, on the other. The existence of synaptic transmission between hair cells and afferent nerve fibers is a characteristic feature of all receptors in the acousticolateral system (Roberts et al. 1991). In the evolutionary process, the development of such an organization of the peripheral part of certain sensory systems has a number of advantages over primary receptors. For example, for many reasons it is important to understand the mechanisms of synaptic transmission. At the synaptic level the local electric potentials are transformed into impulse activity in the nerve fibers. In the receptors of the acousticolateral system, the synapses not only transmit signals from receptor cells to afferent fibers but also contribute to an extremely high sensitivity of hair cells, as does the apical membrane. The presence of synaptic transmission makes it possible for a great number of receptor cells to converge on one afferent nerve fiber, which seems essential in an increase of sensory sensitivity.

An important property of hair cells is their ability to release a neurotransmitter permanently, thereby producing a stationary depolarization and impulse activity in afferent nerve fibers. This enables receptors to respond to stimuli of opposite polarity. In this regard, a very important feature of tonic receptors is the practically total absence of desensitization processes in the postsynaptic receptors. This property may be significant for a tonic receptor to maintain high sensitivity in the face of continued concentration of the neurotransmitter.

During the past two decades, great achievments have been made in the investigation of transduction processes in both primary and secondary mechanoreceptors (Akoev and Alekseev 1985; Akoev et al. 1988b). It is noteworthy that kinocilium is not required for the transduction process in vertebrate hair cells. The transduction channels are located at the distal part of stereocilia. At rest, these channels are active, and the inward receptor current is 10–20% of its maximal level. This resting activation permits a hair cell to respond to both movement directions of the hair bundle. That is why the hair cell transduction process has no absolute threshold. These transduction channels are nonspecific in nature, and a wide variety of mono- and divalent cations can pass through them. The transduction channels are not voltage-sensitive, and there are several channels per stereocilium. The receptor current in vivo results mainly from K^+ flow. Voltage-clamp experiments on isolated hair cells revealed the existence of inward and outward currents. During cell depolarization, the inward Ca^{2+} current is activated, resulting in the involvement of neurotransmitter release. Besides this current, depolarization of hair cells also evokes two outward K^+ currents, namely, transitory current and Ca^{2+}-activated K^+ conductance. The existence of several types of conductances permits hair cells to analyze external stimuli by means of resonance frequency.

In afferent fibers of the acousticolateral system, spontaneous and evoked excitatory postsynaptic potentials are recorded. Hair cells are thought to possess many transmitter release sites with different sensitivity. This permits the hair cell to respond to natural stimuli in a wide range of intensities. Ca^{2+} is essential in transmitter release. Electrical processes in hair cells are regulated by efferent fibers. It has been established that stimulation of efferent fibers evokes inhibitory postsynaptic potentials. During efferent stimulation, hair cell sensitivity to mechanical stimulation is 25 times less. It should be noted that in some receptors of the acousticolateral system (electroreceptors and vestibular apparatus in lower vertebrates), efferent innervation is lacking.

The available data provide evidence that dicarboxylic amino acids are the most probable neurotransmitters in the receptors of the acousticolateral system. The identities of the transmitter and its receptors are still unknown. One may suppose that there is only one neurotransmitter in receptors of the acousticolateral system, and that it acts through different subtypes of postsynaptic receptors in various organs. Further studies on the molecular mechanisms of sensory transduction in the hair cells are necessary to clarify many unsolved problems.

References

Adam-Vizi V (1992) External Ca^{2+}-independent release of neurotransmitters. J Neurochem 58: 395–405

Adams DJ, Gillespie JI (1988) The actions of L-glutamate at the postsynaptic membrane of the squid giant synapse. J Exp Biol 140: 535–548

Adams JC, Mroz EA, Sewell WF (1987) A possible neurotransmitter role for CGRP in a hair-cell sensory organ. Brain Res 419: 347–351

Ades AW, Engström H (1974) Anatomy of the inner ear. In: Keidel WD, Nef WD (eds) Auditory system. Springer, Berlin Heidelberg New York, pp 125–158 (Handbook of sensory physiology, vol V/1)

Akoev GN (1982) The effect of Mg^{2+} and Ca^{2+} on the excitability of Pacinian corpuscle. Brain Res 239: 391–399

Akoev GN (1990) Temperature sensitivity of the ampullae of Lorenzini of elasmobranchs. In: Bligh J, Voigt K (eds) Thermoreception and temperature regulation. Springer, Berlin Heidelberg New York, pp 44–52

Akoev GN, Alekseev NP (1985) Functional organization of mechanoreceptors. Nauka, Leningrad (in Russian)

Akoev GN, Andrianov GN (1981) The action of divalent ions and drugs on thermal and electrical sensitivity of the ampullae of Lorenzini. In: Szabo T, Czeh G (eds) Sensory physiology of aquatic lower vertebrates. Pergamon, Oxford Akademiai Kiado, Budapest, pp 57–73 (Advances in physiological science, vol 31)

Akoev GN, Andrianov GN (1988) Synaptic transmission in acousticolateral receptors. Adv Physiol Sci USSR 19: 81–98 (in Russian)

Akoev GN, Andrianov GN (1989) Synaptic transmission in the mechano- and electroreceptors of the acousticolateral system. Prog Sens Physiol 9: 53–95

Akoev GN, Andrianov GN, Ryzova IV, Sherman NO (1992) L-serine-o-phosphate blocks the NMDA-induced responses in the ampullae of Lorenzini of the skates. Neuroreport 3: 7–9

Akoev GN, Dambinova SA, Gorodinski AI, Andrianov GN, Ryzova IV, Stiles RN (1991) Characteristics of excitatory amino acid receptors in sensory synapse. Soc Neurosc Abst 17: p 246

Akoev GN, Elman SI (1974) Potassium ions and electrical activity of single mechanoreceptors, Pacinian corpuscle. Sechenov Physiol J USSR 60: 55–61 (in Russian)

Akoev GN, Ilyinsky OB (1972) Some functional features of the electroreceptors of the ampullae of Lorenzini of Black Sea skates. Proc Acad Sci USSR 205: 499–501 (in Russian)

Akoev GN, Ilyinsky OB (1973) Some functional characteristics of the electroreceptors (the ampullae of Lorenzini) of elasmobranchs. Experientia 29: 293–294

Akoev GN, Muraveiko VM (1984) Physiological properties of lateral-line receptors of the lamprey. Neurosci Lett 49: 171–173

Akoev GN, Volpe NO (1978) Analysis of the mechanism of thermal sensitivity of the ampullae of Lorenzini of the skate. Sechenov Physiol J USSR 64: 1681–1688 (in Russian)

Akoev GN, Zhadan GG (1978) Perception of chemical and thermal stimuli by the ampullae of Lorenzini in skates. J Evol Biochem Physiol 14: 266–272 (in Russian)

Akoev GN, Alekseev NP, Ilyinsky OB (1971) The modern view on the electroreception. Prog Physiol Sci 2: 122–142 (in Russian)
Akoev GN, Zhadan PM, Ilyinsky OB, Titkov ES (1974) Electrophysiological investigation of the electroreceptors (the ampullae of Lorenzini) in Black Sea skates. Neurophysiology (USSR) 6: 426–433 (in Russian)
Akoev GN, Beller NN, Zhadan PM, Ilyinsky OB, Titkov ES (1975) Some functional characteristics of electroreceptors of electric and nonelectric fish. Sechenov Physiol J USSR 61: 391–399 (in Russian)
Akoev GN, Ilyinsky OB, Zhadan PM (1976a) Physiological properties of electroreceptors of marine skates. Comp Biochem Physiol 53(A): 201–209
Akoev GN, Ilyinsky OB, Zhadan PM (1976b) Responses of electroreceptors (ampullae of Lorenzini) of skates to electric and magnetic fields. J Comp Physiol 106: 127–136
Akoev GN, Andrianov GN, Volpe NO (1980a) Effect of calcium ions on electrical and thermal sensitivity of the ampullae of Lorenzini in Black Sea skates. Sechenov Physiol J USSR 66: 56–63 (in Russian)
Akoev GN, Andrianov GN, Volpe NO (1980b) L-glutamate as possible neurotransmitter in the ampullae of Lorenzini of the skate. Neurosci Lett 20: 307–312
Akoev GN, Makovsky VS, Volpe NO (1980c) Effects of tetraethylammonium on mechano- and electrosensitive channels of Pacinian corpuscle. Neurosci Lett 19: 61–64
Akoev GN, Volpe NO, Zhadan GG (1980d) Analysis of effects of chemical and thermal stimuli on the ampullae of Lorenzini of the skates. Comp Biochem Physiol 65(A): 193–201
Akoev GN, Andrianov GN, Volpe NO (1981) A study of the mechanisms of synaptic transmission in the skate ampullae of Lorenzini. Neurophysiology (USSR) 13: 292–298 (in Russian)
Akoev GN, Andrianov GN, Volpe NO (1982) Effect of acetylcholine and cholinolytics on synaptic processes in ampullae of Lorenzini of the skates. Proc Acad Sci USSR 262: 733–736 (in Russian)
Akoev GN, Andrianov GN, Sherman NO (1983) Effect of acetylcholine and cholinolytics on the synaptic transmission in the ampullae of Lorenzini of the skate. Comp Biochem Physiol 74(C): 95–97
Akoev GN, Andrianov GN, Sherman NO (1984a) Effect of potassium ions on the impulse activity of the ampullae of Lorenzini in skates. Neurophysiology (USSR) 16: 493–497 (in Russian)
Akoev GN, Andrianov GN, Sherman NO (1984b) Effects of potassium ions on the impulse activity of the ampullae of Lorenzini of the skate. Neurosci Lett 46: 341–345
Akoev GN, Andrianov GN, Sherman NO (1984c) Effects of glutamic and aspartic amino acids on the impulse activity of the ampullae of Lorenzini in skates. Sechenov Physiol J USSR 70: 322–325 (in Russian)
Akoev GN, Andrianov GN, Sherman NO (1985) Effects of GABA, taurine and catecholamines on synaptic processes in the ampullae of Lorenzini of Black Sea skates. Sechenov Physiol J USSR 71: 466–471 (in Russian)
Akoev GN, Andrianov GN, Sherman NO (1986a) The action of kainic acid on the synaptic transmission in electroreceptors (ampullae of Lorenzini) of skate. Neurophysiology (USSR) 18: 147–153 (in Russian)
Akoev GN, Andrianov GN, Jakovleva MF, Kuznetsov VI, Moiseeva AB, Ryzhova IV, Sherman NO (1986b) Studies of the mechanism of synaptic transmission in electroreceptors (ampullae of Lorenzini) in skates. Proc IXth Meeting Evol Physiol. Nauka, Leningrad, p 8 (in Russian)
Akoev GN, Andrianov GN, Szabo T (1987a) Effect of kainic acid on synaptic transmission in electroreceptors (the ampullae of Lorenzini) of skates. Comp Biochem Physiol 88(A): 61–63
Akoev GN, Andrianov GN, Sherman NO (1987b) Effects of aspartic acid in the ampullary electroreceptors of skates. Neurophysiology (USSR) 19: 61–67 (in Russian)
Akoev GN, Andrianov GN, Sherman NO (1987c) Analysis of the effect of glutamic acid diethylester on synaptic transmission in the ampullae of Lorenzini of skates. Neurophysiology (USSR) 19: 323–327 (in Russian)

Akoev GN, Andrianov GN, Sherman NO (1988a) Effects of some amino acid antagonists on synaptic transmission in the electroreceptors (the ampullae of Lorenzini) of skates. Neurophysiology (USSR) 20: 457–463 (in Russian)
Akoev GN, Alekseev NP, Krylov BV (1988b) Mechanoreceptors. Their functional organization. Springer, Berlin Heidelberg New York
Akoev GN, Andrianov GN, Bromm B, Szabo T, Sherman NO (1989) Analysis of the effect of quisqualate, N-methyl-D-aspartate and some amino acid antagonists on synaptic transmission in the ampullae of Lorenzini. Neurophysiology (USSR) 21: 160–167 (in Russian)
Akoev GN, Andrianov GN, Bromm B, Szabo T (1991a) Neuropharmacological analysis of transmission in afferent synapses of electroreceptors (ampullae of Lorenzini) in the marine skate, *Raja clavata*. J Comp Physiol 168: 639–646
Akoev GN, Andrianov GN, Szabo T, Bromm B (1991b) Effects of excitatory amino acid antagonists on synaptic transmission in the ampullae of Lorenzini of the skate, *Raja clavata*. J Comp Physiol 168: 647–652
Altman H, Ten Bruggencate G, Pickelmann P, Steinberg R (1976) Effects of glutamate, aspartate and two presumed antagonists on feline rubrospinal neurones. Pfluegers Arch 364: 249–255
Altschuler RA, Neises GR, Harmison GG, Wenthold RJ, Fex J (1981) Immunocytochemical localization of aspartate aminotransferase immunoreactivity in the cochlear nucleus of the guinea pig. Proc Natl Acad Sci USA 78: 6553–6557
Altschuler RA, Parakkal MH, Rubio JA, Hoffman DW, Fex J (1984) Enkephalin-like immunoreactivity in the guinea pig organ of Corti: ultrastructural and lesion studies. Hear Res 16: 17–31
Altschuler RA, Kachar B, Rubio JA, Parakkal MH, Fex J (1985) Immunocytochemical localization of choline acetyltransferase-like immunoreactivity in the guinea pig cochlea. Brain Res 338: 1–11
Altschuler RA, Sherid CE, Horn JW, Wenthold RJ (1989) Immunicytochemical localization of glutamate immunoreactivity in the guinea pig cochlea. Hear Res 42: 167–174
Andres KH, von During M (1988) Comparative anatomy of vertebrate electroreceptors. Prog Brain Res 74: 113–131
Andres KH, von During M, Petrasch E (1988) The fine structure of ampullary and tuberous electroreceptors in the South American blind catfish *Pseudocetopsis* sp. Anat Embryol 177: 523–535
Andrianov GN, Ilyinsky OB (1983) Lateral-line organs. In: Kreps EM (ed) Evolutional physiology. Nauka, Leningrad, pp 110–160 (Handbook of physiology, vol 2) (in Russian)
Andrianov GN, Brown HR, Ilyinsky OB (1974) Responses of central neurons on electrical and magnetic stimuli of the ampullae of Lorenzini in the Black Sea skate. J Comp Physiol 93: 287–299
Andrianov GN, Brown HR, Ilyinsky OB (1979) A review: physiology of the fish electroreceptors. Sechenov Physiol J USSR 65: 785–800 (in Russian)
Andrianov GN, Bretschneider F, Peters RC, Teunis PFM (1992a) In vitro electroreceptor organs for pharmacological studies. J Neurosci Meth 44: 1–6
Andrianov GN, Bretschneider F, Peters RC (1992b) The actions of L-glutamate and its agonists on the ampullary electroreceptor organs of the catfish Ictalurus nebulosus. Comp Biochem Physiol 103C: 65–71
Andrianov GN, Akoev GN, Braun HA, Ryzhova IV, Voigt KH, Sherman NO (1993) Actions of dalargin upon single unit activity in the ampullae of Lorenzini of the skate Raja clavata. NeuroReport 4: 53–54
Annoni J-M, Cochran SL, Precht W (1984) Pharmacology of the vestibular hair cell-afferent fiber synapse in the frog. J Neurosci 4: 2106–2116
Ariëns Kappers CU, Huber GC, Crosby EC (1936) The comparative anatomy of the nervous system of vertebrates including man, vol. MacMillan, New York
Art JJ, Fettiplace R (1987) Variation of membrane properties in hair cells isolated from the turtle cochlea. J Physiol (Lond) 385: 207–242
Art JJ, Crowford AC, Fettiplace R, Fuchs PA (1982) Efferent regulation of hair cells in the turtle cochlea. Proc R Soc Lond B 216: 377–384

Art JJ, Crowford AC, Fettiplace R (1986) Electrical resonance and membrane currents in turtle cochlear hair cells. Hear Res 22: 31–36
Ascher P, Nowak L (1986) Calcium permeability of the channels activated by N-methyl-D-aspartate (NMDA) in isolated mouse central neurons. J Physiol (Lond) 377: 35 P
Ascher P, Nowak L (1987) Electrophysiological studies of NMDA receptors. Trends Neurosci 10: 284–288
Ashmore JF (1987) A fast motile response in guinea pig outer hair cells: the cellular basis of the cochlear amplifier. J Physiol (Lond) 338: 323–347
Ashmore JF (1988a) Hair cells. Sci Prog 72: 139–153
Ashmore JF (1988b) Ionic mechanisms in hair cells of the mammalian cochlea. Prog Brain Res 74: 3–9
Ashmore JE (1990) Mechanoreception. In: Borsellino A, Cervetto L, Torre V (eds) Sensory transduction. Plenum Press, New York, pp 25–50
Ashmore JF (1991) The electrophysiology of hair cells. Annu Rev Physiol 53: 465–476
Ashmore JF (1992) Mammalian hearing and the cochlea cellular mechanisms of the cochlear amplifier. In: Corey DP, Roper SD (eds) Sensory transduction. Soc gen physiol 45th annual symp. Rockfeller Univ Press, pp 395–412
Ashmore JF, Meech RW (1986) Ionic basis of membrane potential in outer hair cells of guinea pig cochlea. Nature 322: 368–371
Ashmore JF, Ohmori H (1990) Control of intracellular calcium by ATP in isolated outer hair cells of the guinea-pig cochlea. J Physiol (Lond) 428: 109–131
Assad JA, Corey DP (1992) An active motor model for adaptation by vertebrate hair cells. J Neurosci 12: 3291–3309
Assad JA, Hacohen N, Corey DP (1989) Voltage dependence of adaptation and active bundle movements in bullfrog saccular hair cells. Proc Natl Acad Sci USA 86: 2918–2922
Assad JA, Shepherd GMG, Corey DP (1991) Tip-link integrity and mechanical transduction in vertebrate hair cells. Neuron 7: 985–994
Ault B, Evans RH, Frankis AA, Oakes DJ, Watkins JC (1980) Selective depression of excitatory amino acid induced depolarizations by magnesium ions in isolated spinal cord preparations. J Physiol (Lond) 307: 413–428
Baden-Kristensen K, Weiss TF (1983) Receptor potentials of lizard hair cells with free-standing stereocilia: responses to acoustic clicks. J Physiol (Lond) 355: 699–721
Bagger-Sjöbäck D (1974) The sensory hairs and their attachments in the lizard basilar papilla. Brain Behav Evol 10: 88–94
Baird IL (1974) Anatomical features of the inner ear in submammalian vertebrates. In: Keidel WD, Neff WD (eds) Auditory system. Anatomy, physiology (ear). Springer, Berlin Heidelberg New York, pp 159–212 (Handbook of Sensory Physiology, vol V/1)
Balcar VJ, Johnston GAR (1972) Glutamate uptake by brain slices and its relation to the depolarization of neurones by acidic amino acids. J Neurobiol 13: 295–301
Balcar VJ, Johnston GAR, Twitchin B (1977) Stereospecificity of the inhibition of L-glutamate and L-aspartate high affinity uptake in rat brain slices by threo-3-hydroxyaspartate. J Neurochem 28: 1145–1146
Barets A, Szabo T (1962) Appareil synaptique des cellules sensorielles de l'ampoule de Lorenzini chez la torpille, *Torpedo marmorata*. J Microsc (Paris) 1: 47–54
Barets A, Szabo T (1964) Ultrastructure des cellules sensorielles des mormyromastes de *Gnathonemus petersii* et de leur appareil synaptique. J Microsc (Paris) 3: 85–90
Barrett EF, Barrett JN, Botz D, Chang DB, Mahaffey (1978) Temperature-sensitive aspects of evoked and spontaneous transmitter release at the frog neuromuscular junction. J Physiol (Lond) 279: 253–273
Barron SE, Bobbin RP, Guth P, Norris C (1987) Cytochalasin D suppresses sound evoked potentials in the guinea pig cochlea. Hear Res 31: 147–154
Baumann M, Roth A (1986) The Ca^{2+} permeability of the apical membrane in neuromast hair cells. J Comp Physiol (A) 158: 681–688
Bauswein E (1977) Effect of calcium on the differentiating operation of the ampullary electroreceptor in *Ictalurus nebulosus*. J Comp Physiol (A) 121: 381–394
Bennett MVL (1967) Mechanisms of electroreception. In: Cahn P (ed) Lateral line detectors. Indiana Univ Press, Bloomington, pp 331–393

Bennett MVL (1971) Electroreception. In: Hoar WS, Randell DJ (eds) Fish physiology. Academic Press, New York, pp 493–574
Bennett MVL (1976) Transmission at receptor synapses. In: Desiraju T (ed) Mechanisms in transmission of signal for conscious behavior. Elsevier, Amsterdam, pp 345–366
Bennett MVL, Obara S (1986) Ionic mechanisms and pharmacology of electroreceptors. In: Bullock TH, Heiligenberg W (eds) Electroreception. Wiley, New York, pp 157–181
Bennett MVL, Sandri C, Akert K (1989) Fine structure of the tuberous electroreceptor of the high-frequency electric fish, *Sternarchus albifrons* (Gymnotiformes). J Neurocytol 18: 265–283
Benshalom G, Flock Å (1977) Calcium-induced density in synaptic vesicles of afferent and efferent synapses on hair cells in the lateral line organs. Brain Res 121: 173–178
Benshalom G, Flock Å (1980) Ultrastructure of synapses in the lateral line canal organ. Acta Oto-Laryngol 90: 101–174
Bernard C (1983) Action of streptomycin and calcium on the apical membrane of hair cells of the frog isolated semicircular canal. Acta Oto-Laryngol 96: 21–30
Bernard C, Cochran SL, Precht W (1983) Presynaptic action of cholinergic agents upon the hair cell-afferent fiber synapse in the vestibular labyrinth of frog. Brain Res 338: 225–236
Biscoe TJ, Evans RH, Headley PM, Davies J, Dray A, Francis AA, Martin MR, Watkins JC (1976) Structure-activity relations of excitatory amino acids in frog and rat spinal neurones. Br J Pharmacol 58: 373–382
Biscoe TJ, Davies J, Dray A, Evans RH, Francis AA, Martin MR, Watkins JC (1977a) Depression of synaptic excitation and of amino acid-induced excitatory responses by D-aminoadipate, -diaminopimelic acid and HA-966. Eur J Pharmacol 45: 315–316
Biscoe TJ, Evans RH, Francis AA, Martin JC, Davis J, Dray A (1977b) D-aminoadipate as a selective antagonist of amino acid-induced and synaptic excitation of mammalian spinal neurons. Nature 270: 743–745
Blagoeva P, Masi I, Scoti De Carolis A, Longo VG (1972) Amino acid release from the cerebral cortex of the rabbit, its relationship with the electrocorticogramme. Physiol Behav 9: 307–313
Bledsoe SC Jr (1986) Pharmacology and neurotransmission of sensory transduction in the inner ear. In: Miller JM (ed) Seminars in hearing, vol 7. Thieme, New York, pp 117–138
Bledsoe SC Jr, Bobbin RP (1982) Effects of D-aminoadipate on excitation of afferent fibres in the lateral line of *Xenopus laevis*. Neurosci Lett 32: 315–320
Bledsoe SC Jr, Bobbin RP, Thalman R, Thalman I (1980) Stimulusinduced release of endogenous amino acids from skins containing the lateral line organ of *Xenopus laevis*. Exp Brain Res 40: 97–101
Bledsoe SC Jr, Bobbin RP, Chihal DM (1981a) Kainic acid: an evaluation of its action on cochlear potentials. Hear Res 4: 109–120
Bledsoe SC Jr, Bobbin RP, Chihal DM (1981b) Technique for studying sound-induced release of endogenous amino acids from guinea pig cochlea. Assoc for Res in Otolaryngol, St Petersburg Beach, Florida, p 27
Bledsoe SC Jr, Chihal DM, Bobbin RP, Morgan DN (1983) Comparative actions of glutamate and related substances on the lateral line of *Xenopus laevis*. Comp Biochem Physiol (C) 75: 199–206
Bledsoe SC Jr, Meyer JR, Howland MM (1985) Effects of excitatory amino acid receptor antagonists on synaptic excitation in the lateral line of *Xenopus laevis*. Assoc for Res in Otolaryngol, Clearwater Beach, Florida, p 111
Bledsoe SC Jr, Bobbin RP, Puel JL (1988) Neurotransmission in the inner ear. In: Jahn AF, Santos-Sacchi J (eds) Physiology of the ear. Raven, New York, pp 385–406
Bledsoe SC Jr, McLaren JD, Meyer JR (1989a) Potassium-induced release of endogenous glutamate and two as yet unidentified substances from the lateral line of *Xenopus laevis*. Brain Res 493: 113–122
Bledsoe SC Jr, Sinard RJ, Allen SJ (1989b) Analysis of histamine as a hair-cell transmitter in the lateral line of *Xenopus laevis*. Hear Res 38: 81–94

Bobbin RP (1979) Glutamate and aspartate mimic the afferent transmitter in the cochlea. Exp Brain Res 34: 389–393

Bobbin RP, Guth RS (1970) Evidence that gamma-aminobutyric acid is not the inhibitory transmitter at the crossed olivo-cochlear nerve-hair cell junction. Neuropharmacology 9: 567–574

Bobbin RP, Konishi T (1974) Action of cholinergic and anticholinergic drugs at the crossed olivo-cochlear bundle-hair cell junction. Acta Oto-Laryngol 77: 56–65

Bobbin RP, Thompson MH (1978) Effects of putative transmitters on afferent cochlear transmission. Ann Otol Rhinol Laryngol 87: 185–190

Bobbin RP, Bledsoe SC Jr, Chihal DM (1981a) Effects of various excitatory amino acid antagonists on guinea pig cochlear potentials. Assoc for Res in Otolaryngol, St Petersburg Beach, Florida, p 27

Bobbin RP, Bledsoe SC Jr, Chihal DM, Morgan DN (1981b) Comparative actions of glutamate and related substances on the *Xenopus laevis* lateral line. Comp Biochem Physiol (C) 69: 145–147

Bobbin RP, Bledsoe SC Jr, Jenison GL, Winberg S, Caesar G (1983) Actions of atropine and bicuculine on the activity of afferent fibers in the *Xenopus laevis* lateral line. Soc Neurosci Abstr 9: p 739

Bobbin RP, Bledsoe SC Jr, Jenison GL (1984) Neurotransmitters of the cochlea and lateral line organs. In: Berlin CJ (ed) Recent advances in hearing science. College-Hill Press, San Diego CA, pp 159–180

Bobbin RP, Ceasar G (1987) Kynurenic acid and gamma-D-glutamylaminomethylsulfonic acid suppress the compound action potential of the auditory nerve. Hear Res 25: 77–81

Bobbin RP, Ceasar G, Fallon M (1990) Potassium induced release of GABA and other substances from the guinea pig cochlea. Hear Res 46: 83–94

Bobbin RP, Jastreboff PJ, Fallon M, Littman T (1990) Nimodipine, an L-channel Ca^{2+} antagonist, reverses the negative summating potential recorded from the guinea-pig cochlea. Hear Res 46: 277–288

Bobbin RP, Ceasar G, Fallon M (1991) Changing cation levels (Mg^{2+}, Ca^{2+}, Na^{+}) alters the release of glutamate, GABA and other substances from guinea pig cochlea. Hear Res 54: 135–144

Bodznick D, Northcutt RG (1981) Electroreception in lampreys: evidence that the earliest vertebrates were electroreceptive. Science 212: 465–467

Borg E, Densert O, Flock Å (1974) Synaptic vesicles in the cochlea. Acta Oto-Laryngol 78: 321–332

Bracho H, Budelli R (1978) The generation of resting membrane potentials in an inner ear hair cell system. J Physiol (Lond) 281: 445–465

Bredberg G, Ades HW, Engström H (1972) Scanning electron microscopy of the normal and pathologically altered organ of Corti. Acta Oto-Laryngol Suppl 301: 1–48

Bromm B, Hensel H, Nier K (1975) Response of the ampullae of Lorenzini to static combined electric and thermal stimuli in *Sciliorhinus canicula*. Experientia 31: 615–618

Bromm B, Hensel H, Tagmat AT (1976) The electrosensitivity of the isolated ampullae of Lorenzini in the dogfish. J Comp Physiol 111: 127–136

Brown HR, Govardovsky VI (1981) Skin mechanism of electroreceptor thermal sensitivity of the ampullae of Lorenzini. Neurophysiology (USSR) 13: 307–314 (in Russian)

Brown GR, Govardovsky VI (1982) Changes of transepithelial potential and impulse reactions evoked by thermal stimulation of the ampullae of Lorenzini. Neurophysiology (USSR) 14: 11–18 (in Russian)

Brown HR, Govardovsky VI (1983a) Receptor mechanisms of the electroreceptors (ampullae of Lorenzini) in skates. Neurophysiology (USSR) 15: 178–185 (in Russian)

Brown HR, Govardovsky VI (1983b) Electrical model of the receptor of the ampullae of Lorenzini. Neurophysiology (USSR) 15: 235–241 (in Russian)

Brown HR, Ilyinsky OB (1978) The ampullae of Lorenzini in the magnetic fields. J Comp Physiol 126: 333–341

Brown HR, Ilyinsky OB (1984) Physiology of electroreceptors. Nauka, Leningrad (in Russian)

Brown HR, Ilyinsky OB, Muraveiko VM, Gorshkov ES, Fonarev GA (1979) Evidence that geomagnetic variations can be detected by Lorenzinian ampullae. Nature 277: 648–649
Brown HR, Govardovsky VI, Zhadan GG (1980) A study of the mechanism of temperature sensitivity of the ampullae of Lorenzini. Neurophysiology (USSR) 12: 67–74 (in Russian)
Brown HR, Govardovsky VI, Cherepnov VL (1985) Action of calcium and potassium channel blockers on changes in transepithelial potential and spike responses of Lorenzinian ampullae in the Black Sea skate. Neurophysiology (USSR) 17: 652–660 (in Russian)
Brown MC, Nuttall AL, Masta RI (1983) Intracellular recordings from cochlear inner hair cells: Effects of stimulation of the crossed olivocochlear efferents. Science 222: 69–72
Brownell WE, Bader CR, Bertrand D, De Ribaupierre Y (1985) Evoked mechanical responses of isolated cochlear hair cells. Science 227: 194–196
Bryant GM, Barron SE, Norris CH, Guth PS (1987) Adenosine is a modulator of hair cell afferent neurotransmission. Hear Res 30: 231–238
Bullock TH (1982) Electroreception. Annu Rev Neurosci 5: 121–170
Bullock TH, Heiligenberg W (eds) (1986) Electroreception. Wiley, New York
Bullock TH, Bodznick DA, Northcutt RC (1983) The phylogenetic distribution of electroreception: evidence for convergent evolution of a primitive vertebrate sense modality. Brain Res Rev 6: 25–46
Carpenter DO (1970) Membrane potential produced directly by the Na^+ pump in *Aplysia* neurons. Comp Biochem Physiol 35: 371–385
Carter J, Savasta M, Fage D, Scatton B (1986) 2-oxo-[^{14}C]glutarate is taken up by glutamatergic nerve terminals in the rat striatum. Neurosci Lett 72: 227–231
Chabbert C, Deyau G, Sladeczek F, Lehouelleur J, Sans A (1991) Intracellular free calcium in isolated vestibular hair cells and potassium iontophoresis. NeuroReport 2: 243–246
Chizhmakov IV, Kiskin NI, Krishtal OA, Tsyndrenko AYa (1989) Glycine action on N-methyl-D-aspartate receptors in rat hippocampal neurons. Neurosci Lett 99: 131–136
Churchill JA, Schukhnecht HF, Doran R (1956) Acetylcholineesterase activity in the cochlea. Laryngoscope 66: 1–15
Clusin WT, Bennett MVL (1977a) Calcium-activated conductance in skate electroreceptors. Current clamp experiments. J Gen Physiol 69: 121–143
Clusin WT, Bennett MVL (1977b) Calcium-activated conductance in skate electroreceptors. Voltage clamp experiments. J Gen Physiol 69: 145–182
Clusin WT, Bennett MVL (1979a) The oscillatory responses of skate electroreceptors to small voltage stimuli. J Gen Physiol 73: 685–702
Clusin WT, Bennett MVL (1979b) The ionic basis of oscillatory responses of skate electroreceptors. J Gen Physiol 73: 703–723
Clusin WT, Spray DC, Bennett MVL (1975) Activation of a voltage-insensitive conductance by inward calcium current. Nature 256: 425–427
Collingridge GL, Leiter RAJ (1983) Excitatory amino acid receptors in the vertebrate central nervous system. Pharmacol Rev 40: 143
Collins GGS, Anson J, Surtees L (1983) Presynaptic kainate and N-methyl-D-aspartate receptors regulate excitatory amino acid release in the olfactory cortex. Brain Res 265: 157–159
Comis SD, Leng G (1979) Action of putative neurotransmitters in the guinea pig cochlea. Exp Brain Res 36: 119–128
Cousillas H, Cole KS, Johnstone MB (1988) Effect of spider venom on cochlear nerve activity consistent with glutamatergic transmission at hair-afferent dendrite synapse. Hear Res 36: 213–220
Corey DP, Assad JA (1992) Transduction and adaptation in vertebrate hair cells: correlating structure and function. In: Corey DP, Roper SD (eds) Sensory transduction. Soc gener physiol 45th Annual Sympos Rockfeller Univ Press, pp 325–342
Corey DP, Hudspeth AJ (1979a) Ionic basis of the receptor potential in a vertebrate hair cell. Nature 281: 675–677

Corey DP, Hudspeth AJ (1979b) Response latency of vertebrate hair cells. Biophys J 26: 499–506
Corey DP, Smith WJ, Barres BA, Koroshetz WJ (1987) Calmodulin inhibitors block adaptation in vestibular hair cells. Soc Neurosci Abstr 13: 538
Cornford ME, Barajas L (1977) Cupular secretion by *Xenopus laevis* line organs: autoradiographic evidence for incorporation of ^{3}H-glucose and ^{35}S-sulfate. Am J Anat 69: 115–133
Correia MJ, Christensen BN, Moore LE, Lang DG (1989) Studies of solitary semicircular canal hair cells in the adult pigeon. I. Frequency- and time-domain analysis of active and passive membrane properties. J Neurophysiol 62: 924–934
Cox DWG, Headly PM, Watkins JC (1977) Actions of L- and D-homocysteate in rat CNS: a correlation between low affinity uptake and the time coarses of excitation by microelectrophoretically applied L-glutamate analogues. J Neurochem 29: 579–588
Coyle JI (1983) Neurotoxic action of kainic acid. J Neurochem 41: 1–11
Crawford AC, Evans MG, Fettiplace R (1991) The actions of calcium on the mechano-electrical transducer current of turtle hair cells. J Physiol 434: 369–398
Crawford AC, Fettiplace R (1978) Ringing responses in cochlear hair cells of the turtle. J Physiol (Lond) 284: 120–122
Crawford AC, Fettiplace R (1979) Reversal of hair cell responses by current. J Physiol (Lond) 295: 66
Crawford AC, Fettiplace R (1980) The frequency selectivity of auditory nerve fibres and hair cells in the cochlea of the turtle. J Physiol (Lond) 306: 79–126
Crawford AC, Fettiplace R (1981) Non-linearities in the responses of turtle hair cells. J Physiol (Lond) 315: 317–338
Crawford AC, Fettiplace R (1986) The mechanical properties of ciliary bundles of turtle cochlear hair cells. J Physiol (Lond) 364: 359–379
Crawford AC, Evans MG, Fettiplace R (1989) Activation and adaptation of transducer currents in turtle hair cells. J Physiol (Lond) 419: 405–434
Cull-Candy SG, Usowicz MM (1987) Multiple-conductance channels activated by excitatory amino acids in cerebellar neurons. Nature 325: 525–528
Curtis DR, Watkins JC (1965) The pharmacology of amino acids related to gamma-aminobutyric acid. Pharmacol Rev 17: 347–391
Dallos P (1984) Peripheral mechanisms of hearing. In: Darian-Smith I (ed) Sect 1: The nervous system. Sensory processes, pt 2. Bawerly Press, Bethesda, pp 595–637 (Handbook of Physiology, vol III)
Dallos P (1985a) Response characteristics of mammalian cochlear hair cells. J Neurosci 5: 1591–1608
Dallos P (1985b) Membrane potential and response changes in mammalian cochlear hair cells during intracellular recording. J Neurosci 5: 1609–1615
Dallos P (1986) Neurobiology of cochlear inner and outer hair cells. Intracellular recordings. Her Res 22: 185–198
Dallos DP (1992) The active cochlea. J Neurosci 12: 4575–4585
Dallos P, Cheatham MA (1992) Cochlear hair cells function reflected in intracellular recodings in vivo. In: Corey DP, Roper SD (eds) Sensory transduction. Soc gen physiol 45th annual symp. Rockfeller Univ Press, pp 371–393
Dallos P, Santos-Sacchi J, Flock Å (1982) Intracellular recording from cochlear outer hair cells. Science 218: 582–584
Dambinova SA (1989) Neuroreceptors of glutamate. Nauka, Leningrad (in Russian)
Dambinova SA, Gorodinsky AI (1984) Molecular organization of the glutamate-sensitive chemoexcitatory membranes of nerve cells. L-[^{3}H]glutamate binding to synaptic membrances from rat cerebral cortex. Biochemistry (USSR) 49: 67–74 (in Russian)
Davis H (1965) A model for transducer action in the cochlea. Cold Spring Harbor Symp Quant Biol 30: 181–190
Davis H (1983) An active process in cochlear mechanisms. Hear Res 9: 79–90
Davies J, Watkins JC (1977) Effects of magnesium ions on the responses of spinal neurons to excitatory amino acids and acetylcholine. Brain Res 130: 364–368

Davies J, Watkins JC (1979) Selective antagonism of amino acid-induced and synaptic excitation in the cat spinal cord. J Physiol (Lond) 297: 621–635
Davies J, Evans RH, Francis AA, Watkins IC (1979) Excitatory amino acid receptors and synaptic excitation in the mammalian central nervous system. J Physiol (Paris) 75: 641–645
De Belleroche JS, Bradford HF (1973) Amino acids in synapitc vesicles from mammalian cerebral cortex: a reappraisal. J Neurochem 21: 441–451
Dechesne CJ Raymond J, Sans A (1984) Action of glutamate in the cat labyrinth. Ann Otol Rhinol Laryngol 93: 163–165
Dechesne CJ, Hampson DR, Goping G, Wheaton KD, Wenthold RJ (1991) Identification and localization of a kainate binding protein in the inne ear by electron microscopy immunocytochemistry. Brain Res 545: 223–233
Del Castillo J, Katz B (1954) Changes in endplate activity produced by presynaptic polarization. J Physiol (Lond) 124: 586–604
Dememes D, Wenthold RJ, Moniot B, Sans A (1990) Glatamate-like immunoreactivity in the peripheral vestibular system of mammals. Hear Res 46: 261–270
Denk W, Keolian RM, Webb WW (1992) Mechanical response of frog saccular hair bundles to the aminoglycoside block of mechanoelectrical transduction. J Neurophysiol 68: 927–932
Densert O (1974) Adrenergic innervation in the rabbit cochlea. Acta Oto-Laryngol 78: 345–356
Derbin C (1970) Effects de la section du nerf lateral sur les jonctions sensorineurales des ampoules de Lorenzini de la torpille, *Torpedo marmorata*. J Microsc (Paris) 9: 119–126
Derbin C, Denizot JP, Szabo T (1969) II. Ultrastructure of the type B sense organ of the specific lateral line system of *Gymnarchus niloticus*. Zellforsch 98: 262–276
Desmendt JE, Monaco P (1962) The pharmacology of a centrifugal inhibitory pathway in the cat's acoustic system. Proc First Int Pharmacol Meet 8: 183–188
Dijkgraaf S (1963) The functioning and significance of the lateral line organs. Biol Rev 38: 51–105
Dijkgraaf S, Kalmijn AJ (1966) Versuche zur biologischen Bedeutung der Lorenzinischen Ampullen bei den Elasmobranchirn. Z Vgl Physiol 53: 187–194
Ding JP, Salvi RJ, Sachs F (1991) Stretch-activated ion channels in guinea pig outer hair cells. Hear Res 56: 19–28
Dio K, Mori N, Matsunaga T, Tsumoto T (1990) Blockade of synaptic transmission from hair cells to auditory afferent nerves by 6-cyano-2,3-dihydroxy-7-nitroquinoxaline, a selective non-NMDA receptor antagonist. Eur Arch Oto-Rhino-Laryn 248: 25–30
Disbrow JK, Ruth JA (1985) Differences in L-[^{3}H]glutamate accumulation and endogenous L-glutamate content in synaptic vesicles from mice selectively bred for differences in ethanol sensitivity. J Neurochem 45: 1294–1297
Disler NN (1960) Sense organs of lateral-line system and their importance in fish behavior. Nauka, Moskow (in Russian)
Dohlman G (1971) The attachment of the cupulae, otolith and tectorial membranes to the sensory cell areas. Acta Oto-Laryngol 71: 89–105
Drescher DG, Drescher MJ (1985) HPLC analysis of presumptive neurotransmitters in perilymph. In: Drescher DG (ed) Auditory biochemistry. Charles C Thomas, Springfield, pp 50–67
Drescher DG, Drescher MJ (1987a) Spontaneous neural activity of a mechanoreceptive system is undiminished by replacement of external calcium with equimolar magnesium in the presence of EGTA. Life Sci 40: 1371–1377
Drescher DG, Drescher MJ (1987b) Calcium and magnesium dependence of spontaneous and evoked afferent neural activity in the lateral-line organs of *Xenopus laevis*. Comp Biochem Physiol 87(A): 305–310
Drescher MJ, Drescher DG (1991) N-Acetylhistidine, glutamate, and β-alanine are concentrated in a receptor cell layer of the trout inner ear. J Neurochem 56: 658–664
Drescher MJ, Drescher DG, Medina JE (1983) Effect of sound stimulation at several levels on concentrations of primary amines, including neurotransmitter candidates, in the perilymph of the guinea pig inner ear. J Neurochem 41: 309–320

Drescher MJ, Drescher DG, Hatfield JS (1987) Potassium-evoked release of endogenous primary amine-containing compounds from the trait saccular macula and saccular nerve in vitro. Brain Res 417: 39–50

Drenckhahn D, Kellner J, Mannherz HG, Groschelstewart U, Kendrick-Jones J, Scholey J (1982) Absence of myosin-like immunoreactivity in stereocilia of cochlear hair cells. Nature 300: 531–532

Duggan AW (1974) The differential sensitivity to L-glutamate and L-aspartate of spinal interneurons and Renshow cells. Exp Brain Res 19: 522–528

Dulon D, Mollard P, Aran J-M (1991) Extracellular ATP elevates cytosolic Ca^{2+} in cochlear inner hair cells. NeuroReport 2: 69–72

Dwyer TM, Adams DJ, Hille B (1980) The permeability of the endplate channel to organic cations in frog muscle. J Gen Physiol 75: 469–492

Eatock RA, Corey DP, Hudspeth AJ (1987) Adaptation of mechanoelectrical transduction in hair cells of the bullfrog's sacculus. J Neurosci 7: 2821–2836

Ehrenberger K, Felix D (1990) NMDA receptors are involved in the afferent cochlear neurotransmission in guinea pigs. Abstr Inner Ear Biol 27: 50

Emmerling MR, Sobkowicz HM (1988) Differentiation and distribution of ACh molecular forms in the mouse cochlea. Hear Res 32: 137–146

Engberg I, Flatman JA, Lambert JDC (1979) The actions of excitatory amino acids on motoneurons in the feline spinal cord. J Physiol (Lond) 288: 227–261

Engström H (1958) On the double innervation of the sensory epithelia of the inner ear. Acta Oto-Laryngol 49: 109–118

Engström H (1960) Electron micrographic studies of the receptor cells of the organ of Corti. In: Rasmussen GL, Windle WF (eds) Neural mechanisms of the auditory and vestibular systems. Charles C Thomas, Springfield, Ill, pp 48–62

Engström H, Sjöstrand FS (1954) The structure and innervation of the cochlear hair cells. Acta Oto-Laryngol 44: 490–501

Evans MG, Fuchs PA (1987) Tetrodotoxin-sensitive, voltage-dependent sodium currents in hair cells from the alligator cochlea. Biophys J 52: 649–652

Evans RH, Watkins JC (1978) Specific antagonism of excitant amino acids in the isolated spinal cord of the neonatal rat. Eur J Pharmacol 50: 123–129

Evans RH, Francis AA, Watkins JC (1977) Selective antagonism by Mg^{2+} of amino acid-induced depolarization of spinal neurones. Experientia 33: 489–491

Evans RH, Francis AA, Hunt K, Oake DJ, Watkins JC (1979) Antagonism of excitatory amino acid-induced responses and of synaptic excitation in the isolated spinal cord of the frog. Br J Pharmacol 67: 591–603

Evans RH, Francis AA, Jones AW, Smith DAS, Watkins JC (1982) The effect of series of -phosphonic- -carboxylic amino acids on electrically evoked and excitant amino acid-induced responses in isolated spinal cord preparations. Br J Pharmacol 75: 65–75

Eybalin M, Pujol R (1983) A radioautographic study of [^{3}H] L-glutamate and [^{3}H]-L-glutamine uptake in the guinea pig cochlea. Neuroscience 9: 863–871

Eybalin M, Pujol R (1984) Immunofluorescence with metenkephalin and leuenkephalin antibodies in the guinea pig cochlea. Hear Res 13: 135–140

Eybalin M, Cupo A, Pujol R (1983) Enkephalins in the cochlea. In: Webster WR, Aitkin LM (eds) Mechanism of hearing. Monash Univ Press, Clayton, pp 58–75

Eybalin M, Cupo A, Pujol R (1984) Met-enkephalin characterization in the cochlea: high-performance liquid chromatography and immunoelectron microscopy. Brain Res 305: 313–322

Eybalin M, Abou-Madi L, Rossier J, Pujol R (1985a) Electron microscopic localization of N-terminal pro-enkephalin (synenkephalin) immunostaining in the guinea pig organ of Corti. Brain Res 358: 354–359

Eybalin M, Cupo A, Pujol R (1985b) Met-enkephalin-Arg-Glu-Leu in the organ of Corti: high-performance liquid chromatography and immunoelectron microscopy. Brain Res 331: 389–395

Eybalin M, Pujol R, Bockaert J (1987a) Opioid receptors inhibit the adenylate cyclase in guinea pig cochleas. Brain Res 421: 336–342

Eybalin M, Rebillard G, Jarry T, Cupo A (1987b) Effect of noise level on the met-enkephaline content of the guinea pig cochlea. Brain Res 418: 189–192
Eybalin M, Parnand C, Geffard M, Pujol R (1988) Immunoelectron microscopy identifies several types of GABA-containing efferent synapses in the guinea pig organ of Corti. Neuroscience 24: 29–38
Fagg GE, Lane JD (1979) The uptake and release of putative amino acid neurotransmitters. Neuroscience 4: 1015–1036
Favre D, Sans A (1979) Morphological changes in afferent vestibular hair cell synapses during postnatal development of the cat. J Neurocytol 8: 765–775
Felix D, Ehrenberger K (1977) The action of GABA and acetylcholine in the labyrinth of the cat. In: Portmann M, Aran JM (eds) Inner ear biology. Inserm, Paris, pp 147–157
Felix D, Ehrenberger K (1982) The action of putative neurotransmitter substances in the cat labyrinth. Acta Oto-Laryngol 93: 101–105
Ferrendelli JA, Chang MM, Kinscherf DA (1974) Elevation of cyclic GMP levels in central nervous system by excitatory and inhibitory amino acids. J Neurochem 22: 535–540
Fessard A, Szabo T (1974) Physiology of electroreceptors. In: Fessard A (ed) Electroreceptors and other specialized receptors in lower vertebrates. Springer, Berlin Heidelberg New York, pp 59–124 (Handbook of sensory physiology, vol III/3)
Fettiplace R (1992) The role of calcium in hair cells transduction. In: Corey DP, Roper SD (eds) Sensory transduction. Soc gen physiol 45th annual symp. Rockfeller Univ Press, pp 343–356
Fettiplace R, Crawford AC (1978) The coding of sound pressure and frequency in cochlear hair cells of the terrapin. Proc R Soc B 203: 209–218
Fex J (1973) Neuropharmacology and potentials of the inner ear. In: Møller AR (ed) Basic mechanisms of hearing. Academic Press, New York, pp 377–420
Fex J, Altschuler RA (1981) Enkephalin-like immunoreactivity of olivocochlear nerve fibers in cochlea of guinea pig and cat. Proc Natl Acad Sci USA 78: 1255–1259
Fex J, Altschuler RA (1984) Glutamic acid decarboxylase immunoreactivity of olivocochlear neurons in the organ of Corti of guinea pig and rat. Hear Res 15: 123–131
Fex J, Martin MR (1980) Lack of effect of DL-amino-adipate, an excitatory amino acid antagonist, on cat auditory nerve responses to sound. Neuropharmacology 19: 809–811
Fex J, Wenthold R (1976) Choline acetyltransferase, glutamate decarboxylase and tyrosine hydroxylase in the cochlea and cochlean nucleus of the guinea pig. Brain Res 109: 575–585
Fex J, Fuxe K, Lennerstrand G (1965) Absence of monoamines in olivocochlear fibres in cat. Acta Physiol Scand 64: 259–262
Fex J, Altschuler RA, Wenthold RJ, Parakkal MH (1982) Aspartate aminotransferase immunoreactivity in cochlea of guinea pig. Hear Res 7: 149–160
Fex J, Kachar B, Rubio JA, Parakkal MH, Altschuler RA (1985) Glutamine-like immunoreactivity in cochlea of guinea pig. Hear Res 17: 101–113
Fiedler JL, Inestrosa NG, Bustos G (1986) Putative glutamate receptors in membranes obtained from heads of *Drosophila melanogaster*. J Neurosci Res 16: 505–515
Fields RD, Ellisman MH (1985) Synaptic morphology and differences in sensitivity. Science 228: 197–199
Flock Å (1964) Structure of the macula utriculi with special reference to directional interplay of sensory responses as revealed by morphological polarization. J Cell Biol 22: 413–431
Flock Å (1965) Electron microscopic and electrophysiological studies on the lateral line canal organ. Acta Oto-Laryngol Suppl 199: 1–90
Flock Å (1967) Ultrastructure and function in the lateral line organs. In: Cahn P (ed) Lateral line detectors. Indiana Univ Press, Bloomington, pp 163–197
Flock Å (1971) Sensory transduction in hair cells. In: Loewenstein WR (ed) Principles of receptor physiology. Springer, Berlin Heidelberg New York, pp 396–441 (Handbook of sensory physiology, vol 1)

Flock Å (1974) Information transfer at the synapse between hair cells and sensory nerve fibres. Rheinisch-Westfael Akad Wiss 53: 347–375

Flock Å (1977) Electrone probe determination of relative ion distribution in the inner ear. Acta Oto-Laryngol 83: 239–244

Flock Å, Lam DMK (1974) Neurotransmitter synthesis in inner ear and lateral line organ. Nature 249: 142–144

Flock Å, Russell IJ (1973) The postsynaptic action of efferent fibres in the lateral line organs of the burbot *Lota lota*. J Physiol (Lond) 235: 591–605

Flock Å, Russell IJ (1976) Inhibition of efferent fibres: action on hair cells and afferent synaptic transmission in the lateral line canal organ of turbot, Lota lota. J Physiol (Lond) 257: 45–62

Flock Å, Jørgensen JM, Russell IJ (1973a) The physiology of individual hair cells and their synapses. In: Moller A (ed) Basic mechanism of hearing. Academic Press, New York, pp 273–306

Flock Å, Jørgensen JM, Russell IJ (1973b) Passive electrical properties of hair cells and supporting cells in the lateral line canal organ. Acta Oto-Laryngol 76: 1990–198

Flock Å, Flock B, Myrrey E (1977) Studies on the sensory hairs of receptor cells in the inner ear. Acta Oto-Laryngol 83: 85–91

Flock B, Flock Å, Ulfendahl M (1986) Mechanisms of movement in outer hair cells and a possible structural basis. Acta Oto-Laryngol 234: 83–90

Fonnum F (1984) Glutamate: A neurotransmitter in mammalian brain. J Neurochem 42: 1–11

Foster AC, Fagg GE (1984) Acidic amino acid sites in mammalian neuronal membranes: their characteristics and relationship to synaptic receptors. Brain Res Rev 7: 103–164

Fritzsch B, Munz H (1986) Electroreception in amphibians. In: Bullock TH, Heiligenberg W (eds) Electroreception. Wiley, New York, pp 483–496

Fritzsch B, Wahnschaffe U (1983) The electroreceptive ampullary organs of urodeles. Cell Tissue Res 229: 483–503

Fritzsch B, Wahnschaffe U, Crapon de Caprona MD, Himstedt W (1985) Anatomical evidence for electroreception in larval *Ichthyophis kohtaoensis*. Naturwissenschaften 72: 102–104

Fuchs PA, Evans MG (1988) Voltage oscillations and ionic conductances in hair cells isolated from the alligator cochlea. J Comp Physiol (A) 164: 151–163

Fuchs PA, Nagai T, Evans MG (1988) Electrical tuning in hair cells isolated from chick cochlea. J Neurosci 8: 2460–2467

Fuchs PA, Evans MG (1990) Potassium currents in hair cells isolated from the cochlea of the chick. J Physiol (Lond) 429: 529–551

Fuchs PA, Evans MG, Murrow BW (1990) Calcium currents in hair cells isolated from the cochlea of the chick. J Physiol (Lond) 429: 553–568

Furukawa T (1986) Sound reception and synaptic transmission in goldfish hair cells. Jpn J Physiol 36: 1059–1077

Furukawa T, Ishii Y (1967) Neurophysiological studies on hearing in goldfish. J Neurophysiol 30: 1377–1403

Furukawa T, Matsuura S (1978) Adaptive rundown of excitatory postsynaptic potentials at synapses between hair cells and eighth nerve fibres in the goldfish. J Physiol (Lond) 276: 193–209

Furukawa T, Ishii R, Matsuura S (1972a) An analysis of microphonic potentials of the sacculus of goldfish. Jpn J Physiol 22: 603–616

Furukawa T, Ishii Y, Matsuura S (1972b) Synaptic delay and time course of postsynaptic potentials at the junction between hair cells and eighth nerve fibres in the goldfish. Jpn J Physiol 22: 617–635

Furukawa T, Nayashida Y, Matsuura S (1978) Quantal analysis of the size of excitatory postsynaptic potentials at synapses between hair cells and afferent nerve fibres in goldfish. J Physiol (Lond) 276: 211–226

Furukawa T, Kuno M, Matsuura S (1982) Quantal analysis of a decremental response of hair cell-afferent fibre synapses in the goldfish succulus. J Physiol (Lond) 322: 181–195

Fuxe V (1965) Evidence for the existence of monoamine neurones in the central nervous system. III. The distribution of monoamine nerve terminals in the central nervous system. Acta Physiol Scand Suppl 64: 39–85
Galley N, Klinke R, Oertel W, Pause M, Storch W-H (1973) The effect of intracochleary administered acetylcholine blocking agents on the efferent synapses of the cochlea. Brain Res 64: 55–63
Gannon RP, Laszo CA, Moskovitch DH (1966) The effect of physostigmine on the latency of the cochlear potentials. Acta Oto-Laryngol 61: 536–546
Ganong A, Cotman CW (1982) Acidic amino acid antagonists of lateral perforant path synaptic transmission: agonist-antagonist interactions in the dental gyrus. Neurosci Lett 34: 195–200
Garthwaite J, Balazs R (1979) Supersensitivity to the cyclic GMP response to glutamate during cerebral maturation. Nature 275: 328–329
Gerschenfeld HM (1973) Chemical transmission in invertebrates central nervous systems and neuromuscular junction. Physiol Rev 54: 1–120
Gil-Loyzaga P, Cupo A, Eybalin M (1988) Met-enkephalin and met-enkephalin-Arg-Gly-Leu immunofluorescence in the developing guinea-pig organ of Corti. Dev Brain Res 42: 142–145
Giorguieff MF, Kemel ML, Glowinski J (1977) Presynaptic effect of L-glutamic acid on the release of dophamine in rat striatal slices. Neurosci Lett 6: 73–77
Gisselsson L (1950) Experimental investigation into the problem of humoral transmission in the cochlea. Acta Oto-Laryngol Suppl 82: 3–78
Gleich O, Johnstone BM, Robertson D (1990) Effects of L-glutamate on auditory afferent activity in view of its proposed excitatory transmitter role in the mammalian cochlea. Hear Res 45: 295–312
Gleisner L, Flock Å, Wersäll J (1973) The ultrastructure of the afferent synapse on hair cells in the frog labyrinth. Acta Oto-Laryngol 76: 199–207
Godfrey DA, Carter JA, Berger SJ, Matschinsky FM (1976a) Levels of putative transmitter amino acids in the guinea pig cochlea. J Histochem Cytochem 24: 468–470
Godfrey DA, Krzanovski JJ, Matchinsky FM (1976b) Activities of enzymes of the cholinergic system in the guinea pig cochlea. J Histochem Cytochem 24: 470–472
Godfrey DA, Park JL, Ross CD (1984) Choline acetyltransferase and acetylcholinesterase in centrifugal labyrinthine bundles of rats. Hear Res 14: 93–106
Godukhin OV, Zharikova AD, Novossyolov VI (1979) Release of labelled L-glutamic acid from rat neostriatum on stimulation of frontal cortex in vivo. Sechenov Physiol J USSR 65: 1544–1548 (in Russian)
Godukhin OV, Zharikova AD, Budantsev AYu (1983) Mechanisms of regulation of glutamatergic synaptic transmission in the central nervous system of vertebrates. Adv Mod Biol USSR 95: 383–397 (in Russian)
Görner P (1963) Untersuchungen zur Morphologie und Elektrophylogie des Seitenorgans vom Krallenfrosch (*Xenopus laevis*). Z Vgl Physiol 47: 316–338
Gregory JE, Iggo A, McIntyre AK, Proske U (1988) Receptors in the bill of the platypus. J Physiol (Lond) 400: 349–366
Griguer C, Sans A, Lehouelleur J (1993) Non-typical K^+-current in cesium-loaded guinea pig type I vestibular hair cell. Pflügers Arch 422: 407–409
Guharay F, Sachs F (1984) Stretch activated single ion channel currents in tissue-cultured embryonic chick skeletal muscle. J Physiol (Lond) 352: 685–701
Gulley RL, Reese TS (1977) Freeze-fracture studies on the synapses in the organ of Corti. J Comp Neurol 171: 517–543
Gulley RL, Fex J, Wenthold RJ (1979) Uptake of putative neurotransmitters in the organ of Corti. Acta Oto-Laryngol 88: 177–182
Guth SL, Drescher DG (1990) Effects of divalent cations on the frequency of spontaneous action potentials from the lateral line organ of *Xenopus laevis*. Brain Res 508: 76–84
Guth PS, Melamed B (1982) Neurotransmission in the auditory system: a primer for pharmacologists. Annu Rev Pharmacol Toxicol 22: 383–412
Guth SL, Norris CH (1984) Pharmacology of the isolated semicircular canal: effect of GABA and picrotoxin. Exp Brain Res 56: 72–78

Guth PS, Norris CH, Bobbin RP (1976) The pharmacology of transmission in the peripheral auditory system. Pharmacol Rev 28: 95–125
Guth PS, Sewell WF, Tachibana M (1981) The pharmacology of the cochlear afferents and cochlear nucleus. In: Brown RD (ed) Pharmacology of hearing. Wiley, New York, pp 99–136
Guth PS, Norris CH, Sewell WF (1985) Primary afferent transmission. In: Drescher DG (ed) Auditory biochemistry. Charles C Thomas, Springfield, pp 41–49
Guth PS, Norris CH, Barron SE (1988) Three tests of the hypothesis that glutamate is the sensory hair cell transmitter in the frog semicircular canal. Hear Res 33: 223–228
Guth PS, Aubert A, Ricci AJ, Norris CH (1991) Differential modulation of spontaneous and evoked neurotransmitter release from hair cells: some novel hypotheses. Hear Res 56: 69–78
Hablitz JJ, Langmoen IA (1982) Excitation of hippocampal pyramidal cells by glutamate in the guinea pig and rat. J Physiol (Lond) 325: 317–331
Hacohen A, Assad JA, Smith WJ, Corey DP (1987) Regulation of tension on hair-cell transduction channels: displacement and calcium dependence. J Neurosci 9: 3988–3997
Hacohen N, Assad JA, Smith WJ, Corey DP (1989) Regulation of tension on hair-cell transduction channels: displacement and calcium dependence. J Neurosci 9: 3988–3997
Hama K (1965) Some observations on the fine structure of the lateral line organs of the Japanese Sea eel Lyncorymba nystromi. J Cell Biol 24: 193–210
Hama K, Saito K (1977) Fine structure of the afferent synapse of the hair cells in the saccular macula of the goldfish, with special reference to the anastomosing tubules. J Neurocytol 6: 361–373
Hamilton DW (1968) The caliceal synapse of type I vestibular hair cells. J Ultrastruct Res 23: 98–114
Hammerstad JP, Cutler RWP (1972) Sodium ions movements and sponstaneous and electrically stimulated release of [^{3}H]GABA and [^{14}C]glutamic acid from rat cortical slices. Brain Res 47: 401–413
Hammerstad JP, Murray JE, Cutler RWP (1971) Efflux of amino acid neurotransmitters from rat spinal cord slices. II. Factors influencing the electrically induced efflux of [^{14}C]glycine and [^{3}H]GABA. Brain Res 35: 357–367
Harris GG, Flock Å (1967) Spontaneous and evoked activity from the *Xenopus laevis* lateral line. In: Cahn P (ed) Lateral line detectors. Indiana Univ Press, Bloomington, pp 135–161
Harris GG, Frischkopf LS, Flock Å (1970) Receptor potentials from hair cells of the lateral line. Science 167: 76–79
Harrison NL, Simmonds MA (1985) Quantitative studies on some antagonists of N-methyl-D-aspartate in slices of rat cerebral cortex. Br J Pharmacol 84: 318–391
Hashimoto T, Katsuki Y (1972) Enhancement of the mechanosensitivity of hair cells of the lateral line by environmental potassium ions. J Acoust Soc Am 52: 553–557
Head RA, Tunnicliff G, Matheson GK (1980) Glutamate receptor binding to cat central nervous system membranes. Can J Biochem 58: 534–538
Headly PM, Grillner S (1990) Excitatory amino acids and synaptic transmission: the evidence for a physiological function. Trends Pharmacol Sci 11: 205–211
Hensel H (1955) Quantative Beziehungen zwischen Temperaturreiz und Aktionpotentialen der Lorenzinischen Ampullen. Z Vgl Physiol 37: 509–526
Hensel H (1974) Effect of temporal and spatial temperature gradients on the ampullae of Lorenzini. Pfluegers Arch 347: 89–100
Hensel H, Nier K (1971) Integrated static activity of the ampullae of Lorenzini after long-term exposure to various temperatures. Pfluegers Arch 323: 263–279
Hertz L (1979) Functional interactions between neurons and astrocytes. I. Turnover and metabolism of putative amino acid transmitters. Prog Brain Res 13: 277–323
Higuchi T, Nagai T, Umekita SI, Obara S (1980) The afferent neurotransmitter in the ampullary electroreceptors: L-glutamate mimics the natural transmitter. Neurosci Lett Suppl 4: S7

Hildebrand JG, Townsel JG, Kravitz EA (1974) Distribution of acetylcholine, choline, choline acetyltransferase and acetyltransferase in regions and single identified axons of the lobster nervous system. J Neurochem 23: 951–963
Hilding DA (1964) Catecholamine activity in the inner ear. J Acoust Soc Am 36: 1016–1021
Hillman DE (1972) Observations on morphological features and mechanical properties of the peripheral vestibular receptor system in the frog. Prog Brain Res 37: 70–75
Hillman HH, McIlwain H (1961) Membrane potentials in mammalian cerebral tissue in vitro: dependence on ionic environment. J Physiol (Lond) 157: 263–278
Hoffman DW (1986) Opioid mechanisms in the inner ear. In: Altschuler RA, Hoffman DW, Bobbin RP (eds) The cochlea. Raven, New York, pp 371–382
Hoffman DW, Altschuler RA, Fex J (1983) High-performance liquid chromatographic identification of enkephalin-like peptides in the cochlea. Hear Res 9: 71–79
Hoffman DW, Rubio JA, Altschuler RA, Fex J (1984) Several distinct receptor binding enkephalins in olivocochlear fibers and terminals in the organ of Corti. Brain Res 322: 59–66
Hoffmann DW, Zamir N, Rubio JA (1985) Proenkephaline and prodynorphine-related neuropeptides in the cochlea. Hear Res 17: 47–50
Holley MC, Ashmore JE (1988) On the mechanism of a high-frequency force generator in outer cells isolated from the guinea pig cochlea. Proc R Soc Lond B 232: 413–429
Holton T, Hudspeth AJ (1986a) The transduction channels of hair cells from the bullfrog characterized by noise analysis. J Physiol (Lond) 375: 195–227
Holton T, Hudspeth AJ (1986b) The voltage dependence of haircell transduction. Soc Neurosci Abstr 12: 778
Holton T, Weiss TF (1983) Receptor potentials of lizard cochlear hair cells with free-standing stereocilia in response to tones. J Physiol (Lond) 345: 205–240
Homma S (1983) Physiology and pharmacology of putative transmitters in lamprey central nervous system. Prog Neurobiol 20: 287–311
Housley GD, Norris CH, Guth PS (1988) Histamine and related substances influence neurotransmission in the semicircular canal. Hear Res 35: 87–98
Housley GD, Norris CH, Guth PS (1989) Electrophysiological properties and morphology of hair cells isolated from the semicircular canal of the frog. Hear Res 38: 259–276
Housley GD, Norris CH, Guth PS (1990) Cholinergically induced changes in outward currents in hair cells isolated from the semicircular canal of the frog. Hear Res 43: 121–134
Howard J, Hudspeth AJ (1987) Mechanical relaxation of the hair bundle mediates adaptation in mechanoelectrical transduction by the bullfrog's saccular hair cell. Proc Natl Acad Sci USA 84: 3064–3068
Howard J, Hudspeth AJ (1988) Compliance of the hair bundle associated with gating of mechanoelectrical transduction channels in the bullfrog's saccular hair cell. Neuron 1: 189–199
Howard J, Roberts WM, Hudspeth AJ (1988) Mechano-electrical transduction by hair cells. Annu Rev Biophys Biophys Chem 17: 99–124
Hudspeth AJ (1982) Extracellular current flow and the site of transduction by vertebrate hair cells. J Neurosci 2: 1–10
Hudspeth AJ (1983) Mechanoelectrical transduction by hair cells in the acousticolateralis system. Annu Rev Neurosci 6: 187–215
Hudspeth AJ (1985) The cellular basis of hearing: the biophysics of hair cells. Science 230: 745–752
Hudspeth AJ (1986) The ionic channels of a vertebrate hair cell. Hear Res 22: 21–27
Hudspeth AJ (1989) How the ear's works work. Nature 341: 397–404
Hudspeth AJ (1992) Hair-bundle mechanics and a model for mechanoelectrical transduction by hair cells. In: Corey DP, Roper SD (eds) Sensory transduction. Soc gen physiol 45th annual symp. Rockfeller Univ Press, pp 357–370

Hudspeth AJ, Corey DP (1977) Sensitivity, polarity and conductance change in the response of vertebrate hair cells to controlled mechanical stimuli. Proc Natl Acad Sci USA 74: 2407–2411

Hudspeth AJ, Jacobs R (1979) Stereocilia mediate transduction in vertebrate hair cells. Proc Natl Acad Sci USA 76: 1506–1509

Hudspeth AJ, Lewis RS (1988a) Kinetic analysis of voltage and ion-dependent conductances in saccular hair cells of the bull-frog, *Rana catesbeiana*. J Physiol (Lond) 400: 237–274

Hudspeth AJ, Lewis RS (1988b) A model for electrical resonance and frequency tuning in saccular hair cells of the bull-frog, *Rana catesbeiana*. J Physiol (Lond) 400: 275–297

Igarashi M, Alford BR (1969) Cupula, cupular zone of otolitic membrane and tectorial membrane in the squirrel monkey. Acta Oto-Laryngol 68: 420–426

Ikeda K, Kusakari J, Takasaka T, Saito Y (1987) The Ca^{2+} activity of cochlear endolymph of the guinea pig and the effect of inhibitors. Hear Res 26: 117–125

Ilyinsky OB (1975) Physiology of sensory systems, part 3. Physiology of mechanoreceptors. Handbook of Physiology. Nauka, Leningrad (in Russian)

Ilyinsky OB, Krasnikova TL (1971) On the chemical composition of a medium surrounding some mechano- and electroreceptive structures in fish. J Evol Biochem Physiol 7: 570–575 (in Russian)

Ilyinsky OB, Krasnikova TL (1974) The ionic composition of the jelly of Lorenzini ampullae in Mediterranean electric rays. J Evol Biochem Physiol (USSR) 10: 417–419 (in Russian)

Ilyinsky OB, Chekmasova NH, Cherepnov VL (1974) Structure of ampullary organs of lateral-line system of some elasmobranchs. Proc Acad Sci USSR 210: 1198–1200 (in Russian)

Ishii Y, Matsuura S, Furukawa T (1971) An input-output relation at the synapse between hair cells and eighth nerve fibers in goldfish. Jpn J Physiol 21: 91–98

Iurato S (1961) Submicroscopic structure of the membranous labyrinth. II. The epithelium of Corti's organ. Z Zellforsch 53: 259–298

Iurato S (1962) Efferent fibres to the sensory cells of Corti's organ. Exp Cell Res 27: 162–164

Iurato S (1974) Efferent innervation of the cochlea. In: Keidell D, Neff WD (eds) Auditory system. Springer, Berlin Heidelberg New York, pp 261–282 (Handbook of sensory physiology, vol V/1)

Iurato S, Luciano L, Pannese E, Reale E (1971) Histochemical localization of the acetylcholinesterase (ACHE) activity in the inner ear. Acta Oto-Laryngol Suppl 279: 6–50

Iversen LL, Neal MJ (1968) The uptake of [^{3}H]GABA by slices of rat cerebral cortex. J Neurochem 15: 1141–1149

Jackson H, Parks TN (1989) Spider toxins: recent applications in neurobiology. Annu Rev Neurosci 12: 405–414

Jackson H, Usherwood PNR (1988) Spider toxins as tools for dissecting elements of excitatory amino acid transmission. Trends Neurosci 11: 278–283

Jacobs RA, Hudspeth AJ (1990) Ultrastructural correlates of mechanoelectrical transduction in hair cells of the bullfrog's internal ear. Cold Spring Harbor Symp Quant Biol 55: 547–561

Jahr CE, Stevens CP (1987) Glutamate activates multiple single channel conductances in hippocampal neurons. Nature 325: 522–525

Jahr CE, Yoshioka K (1986) Ia afferent excitation of motoneurons in the in vitro newborn rat spinal cord is selectively antagonized by kynurenate. J Physiol (Lond) 370: 515–530

Jakubovski M (1987) Ultrastructure of the ampullary electroreceptive organs in *Silurus glanis* (Pisces). Z Mikrosk-Anat Forsch (Leipz) 101: 461–472

Jaramillo F, Hudspeth AJ (1991) Localization of the hair cell's transduction channels at the hair bundle's top by iontophoretic application of a channel blocker. Neuron 7: 409–420

Jasser A, Guth PS (1973) The synthesis of acetylcholine on guinea-pig cochlear potentials. J Neurochem 20: 45–53

Jenison GL, Bobbin RP (1986) Quisqualate excites spiral ganglion neurons of the guinea-pig. Hear Res 20: 261–265
Jenison GL, Bobbin RP, Thalman R (1985) Potassium-induced release of endogenous amino acids in the guinea-pig cochlea. J Neurochem 44: 1845–1853
Jenison GL, Winbery S, Bobbin RP (1986) Comparative actions of quisqualate, excitatory amino acid agonists, on guinea-pig cochlear potentials. Comp Biochem Physiol (C) 84: 385–389
Jia M, Nelson PG (1986) Calcium currents and transmitter output in cultured spinal cord and dorsal root ganglion neurons. J Neurophysiol 56: 1257–1267
Johnston GAR (1981) Glutamate uptake and its possible role in neurotransmitter inactivation. In: Roberts PJ, Storm-Mathiesen J, Johnston GAR (eds) Glutamate: transmitter in the central nervous system. Wiley, Chichester, pp 77–87
Johnston GAR, Curtis DR, Davies J, McCulloch RM (1974) Spinal interneuron excitation by conformationally restricted analogues of L-glutamic acid: Nature 248: 804–805
Johnston GAR, Kennedy SME, Twitchin BJ (1979) Action of the neurotoxin kainic acid on high affinity uptake of glutamic acid in rat brain slices. J Neurochem 32: 121–127
Johnston GAR, Lodge D, Bornstein JC, Curtis DR (1980) Potentials of L-glutamate and L-aspartate excitation of cat spinal neurones by the stereoisomers of threo-3-hydroxyaspartate. J Neurochem 34: 241–243
Johnstone BM, Patuzzi R, Syka J, Sykova E (1989) Stimulus-related potassium changes in the organ of Corti of guinea-pig. J Physiol (Lond) 408: 77–92
Jørgensen F (1983) Influence of Ca^{2+} on the mechanosensitivity of the hair cells in the lateral line organs of *Necturus maculosus*. Acta Physiol Scand 118: 423–431
Jørgensen F (1984) Influence of Ca^{2+} on the voltage-dependent mechanosensitivity of lateral line organs of *Xenopus laevis*. Acta Physiol Scand 120: 481–488
Jørgensen F (1988) On the ionic mechanism of hair cells. Prog Brain Res 74: 21–27
Jørgensen F, Ohmori H (1988) Amiloride blocks the mechanoelectric transduction channels of hair cells of the chick. J Physiol (Lond) 403: 577–588
Jørgensen JM (1980) The morphology of the Lorenzinian ampullae of the sturgeon *Acipenser ruthenus* (Pisces: Chondrostei). Acta Zool 61: 87–92
Jørgensen JM (1982) Fine structure of the ampullary organs of the bichir *Polypterus senegalus* Cuvier, 1829 (Pisces: Brachiopterygii) with some notes on the phylogenetic development of electroreceptors. Acta Zool 63: 211–217
Jørgensen JM, Flock Å, (1973) The ultrastructure of lateral line sense organs in the adult salamander *Amblystoma mexicanum*. J Neurophysiol 2: 113–142
Jørgensen JM, Flock Å, Wersäl J (1972) The Lorenzinian ampullae of *Polyodon spathula*. Z Zellforsch 130: 362–377
Junge D, Ortiz CL (1978) Measurement of electrogenic-pump current in *Aplysia* neurones with constant current and constant voltage techniques. J Exp Biol 72: 141–151
Kalmijn AJ (1971) The electric sense of sharks and rays. J Exp Biol 55: 371–392
Kalmijn AJ (1974) The detection of electric fields from inanimate and animate sources other than electric organ. In: Fessard A (ed) Electroreceptors and other specialized receptors in lower vertebrates. Springer, Berlin Heidelberg New York, pp 177–200 (Handbook of sensory physiology, vol III/3)
Kalmijn AJ (1982) Electric and magnetic field detection in elasmobranch fishes. Science 218: 916–918
Karlsen HE, Sand O (1987) Selective and reversible blocking of the lateral line in freshwater fish. J Exp Biol 133: 249–262
Katsuki Y (1973) The ionic receptive mechanism in acoustico-lateralis system. In: Møller AR, Boston P (eds) Basic mechanisms of hearing. Academic Press, New York, pp 307–334
Katz B (1962) The transmission of impulses from nerve to muscle, and the subcellular unit of synaptic action. Proc R Soc Lond B 155: 455–479
Katz B, Miledi R (1967) A study of synaptic transmission in the absence of nerve impulses. J Physiol (Lond) 192: 407–436
Katz RL, Chase TN, Kopin IJ (1969) Effects of ions on stimulus-induced release of amino acids from mammalian brain slices. J Neurochem 16: 961–967

Kawagoe R, Onodera K, Takeuchi A (1981) Release of glutamate from the crayfish neuromuscular junction. J Physiol (Lond) 312: 225–236

Kawagoe R, Onodera K, Takeuchi A (1982) On the quantal release of endogenous glutamate from the crayfish neuromuscular junction. J Physiol (Lond) 322: 529–535

Kawagoe R, Onodera K, Takeuchi A (1984) The uptake and release of the glutamate at the crayfish neuromuscular junction. J Physiol (Lond) 354: 69–78

Kawagoe R, Onodera K, Takeuchi A (1985) Release of endogenous glutamate from the frog spinal cord following dorsal root stimulation. Biomed Res 6: 239–245

Kawagoe R, Onodera K, Takeuchi A (1986) The release of endogenous glutamate from the newborn rat spinal cord induced by dorsal root stimulation and substance P. Biomed Res 7: 253–259

Kawai N, Niwa A, Abe T (1982) Spider venom contains specific receptor blockers of glutaminergic synapses. Brain Res 247: 169–171

Kelly RB, Deutsch JW, Carlson SS, Wagner JA (1979) Biochemistry of neurotransmitter release. Annŭ Rev Neurosci 2: 399–446

Kemp JA, Foster AC, Wong EHF (1987) Non-competitive antagonists of excitatory amino acid receptors. Trends Neurosci 10: 294–299

Kiang NYS, Rho JM, Norhrop CC, Liberman MC, Ryugo DK (1982) Hair cell innervation by spiral ganglion in adult cats. Science 217: 175–177

Kimitsuki T, Ohmori H (1992) The effect of caged calcium release on the adaptation of the transduction current in chick hair cells. J Physiol 458: 27–40

Kimura R (1966) Hairs of cochlear sensory cells and their attachment to the tectorial membrane. Acta Oto-Laryngol 61: 55–72

Kimura R, Wersall J (1962) Termination of the olivocochlear bundle in relation to the outer hair cells of the organ of Corti in guinea pig. Acta Oto-Laryngol 55: 11–32

Kinscherf DA, Chang MM, Ruben EH, Schneider DR, Ferrendelli JA (1976) Comparison of the effects of depolarizing agents and neurotransmitters on regional CNS cyclic GMP levels in various animals. J Neurochem 26: 527–530

Kiskin NI, Tsyndrenko AYa (1985) L-glutamate-activated permeability in the membrane of isolated pyramidal neurons from rat hippocampus. Neurophysiology (USSR) 17: 558–560 (in Russian)

Kiskin NJ, Krishtal OA, Tsyndrenko AYa (1986a) Excitatory amino acid receptors in the membrane of hippocampal pyramidal neurons. Biol Membr (USSR) 3: 909–919 (in Russian)

Kiskin NI, Krishtal OA, Tsyndrenko AYa (1986b) Excitatory amino acid receptors in hippocampal neurons: kainate fails to desensitize them. Neurosci Lett 63: 225–230

Kitajiri M, Yamashita T, Tohyama Y, Kumazawa T, Takeda N, Kawasaki Y, Matsunaga T, Girgis S, Hillyard CJ, MacIntyre I, Emson PC, Shiosaka S, Tohyama M (1985) Localization of calcitonin gene-related peptide in the organ of Corti of the rat: an immunohistochemical study. Brain Res 358: 394–397

Klinke R (1981) Neurotransmitters in the cochlear nucleus. Acta Oto-Laryngol 91: 541–554

Klinke R (1986) Neurotransmission in the inner ear. Hear Res 22: 235–243

Klinke R, Evans EF (1977) Evidence that catecholamines are not the afferent transmitter in the cochlea. Exp Brain Res 21: 315–324

Klinke R, Oertel W (1977a) Amino acids-putative afferent transmitter in the cochlea. Exp Brain Res 30: 145–148

Klinke R, Oertel W (1977b) Evidence that GABA is not the afferent transmitter in the cochlea. Exp Brain Res 28: 311–314

Klinke R, Oertel W (1977c) Evidence that 5-HT is not the afferent transmitter in the cochlea. Exp Brain Res 30: 141–143

Kolesnikov SS, Rebrik TI, Zhainazarov A, Tavartkiladze GA, Kalamkarov GR (1991) A cyclic-AMP-gated conductance in cochlear hair cells. FEBS 290: 167–170

Krawitz EA, Slater CR, Takahashi K, Bownds MD, Grassfeld RJ (1970) Excitatory transmission in invertebrates – glutamate as a potential neuromuscular transmitter compound. In: Andersen P, Jansen JKS (eds) Excitatory synaptic mechanisms. Universitetsforlaget, Oslo, pp 85–93

Krnjevic K (1974) Chemical nature of synaptic transmission in vertebrates. Physiol Rev 54: 418–541
Kroese ABA, van den Bercken J (1980) Dual action of ototoxic antibiotics on sensory hair cells. Nature 283: 395–397
Kroese ABA, van den Bercken J (1982) Effects of ototoxic antibiotics on sensory hair cell functioning. Hear Res 6: 183–197
Kroese ABA, Das A, Hudspeth AJ (1989) Blockade of the transduction channels of hair cells in the bullfrog's sacculus by aminoglycoside antibiotics. Hear Res 37: 203–218
Kros CJ (1990) Electrical properties of the basolateral membrane of hair cells – a review. In: Borsellino A, Cervetto L, Torre V (eds) Sensory transduction. Plenum Press, New York, pp 51–63
Kros CJ, Crawford AC (1990) Potassium currents in inner hair cells isolated from the guinea-pig cochlea. J Physiol (Lond) 421: 263–291
Kros CJ, Rüsch A, Richardson GP (1992) Mechano-electrical transducer currents in hair cells of the cultured neonatal mouse cochlea. Proc Roy Soc B 249: 185–193
Kusakari J, Arakawa E, Rokugo M, Ohyama K, Inamura N (1984) Effect of kainic acid upon Nl latency. Laryngoscope 94: 1365–1369
Kuznetsov VI, Godukhin OB (1985) Mechainism of methionine-enkephaline modulation of glutamatergic transmission in the rat striatum. Neurosci Let 57: 143–146
Kyogoku S, Matsuura S, Kuno M (1986) Generator potentials and spike initiation in auditory fibers of goldfish. J Neurophysiol 55: 244–255
Lambert JDD, Hatman JA, Engberg J (1978) Aspects of the actions of excitatory amino acids on cat spinal motoneurons and on interaction of barbiturate anesthetics. In: Rvall RW, Kelly JC (eds) Iontophoresis and transmitter mechanisms in the mammalian central nervous system. Elsevier, Amsterdam, pp 375–377
Lang DG, Correia MJ (1989) Studies of solitary semicircular canal hair cells in the adult pigeon. II. Voltage-dependent ionic conductances. J Neurophysiol 62: 934–945
Lawrence M (1975) Resting potentials in the inner sulcus and tunnel of Corti. Acta Oto-Laryngol 79: 304–309
Lehtosalo JI, Ylikoski J, Eranko L, Eranko O, Panula P (1984) Immunohistochemical localization of unique enkephalin sequences contained in preproenkephalin A in the guinea pig cochlea. Hear Res 16: 101–107
Leng G (1980) The Davis theory: a review, and implications of recent electrophysiological evidence. Hear Res 3: 17–25
Leng G, Comis S (1979a) Effect of increased potassium in scala tympani on auditory nerve sensitivity. Experientia 35: 767–768
Leng G, Comis S (1979b) The effect of potassium on the activity of auditory nerve fibres on the guinea gig cochlea. Acta Oto-Laryngol 87: 39–46
Levi G, Gallo V, Baiteri M (1980) A reevaluation of veratridine as a tool for studying the depolarization-induced release of neurotransmitters from nerve endings. Neurochem Res 5: 281–295
Lewis ER, Li CW (1975) Hair cell types and distribution in the otolitic and auditory organs of the bullfrog. Brain Res 83: 35–50
Lewis RS, Hudspeth AJ (1983) Voltage- and ion-dependent conductances in solitary vertebrate hair cells. Nature 304: 538–541
Liberman MC (1980) Efferent synapses in the inner hair cell area of the cat cochlea: an electron microscopic study of serial sections. Hear Res 3: 289–304
Liddicoat JD, Roberts BL (1971) The ionic composition of the lateral-line canal fluid of dogfish. J Mar Biol Assoc UK 52: 653–659
Lim DJ (1972) Fine morphology of the tectorial membrane. Arch Oto-Laryngol 96: 199–215
Lim DJ (1986) Functional structure of the organ of Corti: a review. Hear Res 22: 117–146
Lim DJ, Mogi G, O'Dorisio TM, Cataland S (1981) Vasoactive intestinal peptide in perilimph. Assoc for Res in Otolaryngol, St Petersburg Beach, Florida, p 25
Lindeman HH (1969) Studies on the morphology of the sensory regions of the vestibular apparatus. Ergeb Anat Entwicklungsgesch 42: 1–113

Lissmann HW (1958) On the function and evolution of electric organs in fish. J Exp Biol 35: 156–191

Lissmann HW, Machin KE (1958) The mechanism of object location in *Gymnarchus niloticus* and similar fish. J Exp Biol 35: 451–486

Lissman HW, Mullinger AM (1968) Organization of ampullary electric receptors in Gymnotidae (Pisces). Proc R Soc Lond B 169: 345–378

Littman T, Bobbin RP, Fallon M, Puel J-L (1989) The quinoxalinediones DNQX, CNQX and two related congeners suppress hair cell to auditory nerve transmission. Hear Res 40: 45–54

Lodge D, Johnson KH (1990) Noncompetitive excitatory amino acid receptor antagonists. Trends Pharmacol Sci 11: 81–86

Lodge D, Curtis DK, Johnston GAR, Bornstein JC (1980) In vivo inactivation of quisqualate; studies in the cat spinal cord. Brain Res 182: 491–495

Loewenstein WR (1960) Mechanisms of nerve impulse initiation in a pressure receptor (Lorenzinian ampullae). Nature 188: 1034–1035

Loewenstein WR, Ishiko N (1962) Sodium chloride sensitivity and electrochemical effects in a Lorenzinian ampullae. Nature 194: 292–294

Logan WJ, Snyder SH (1972) High affinity uptake systems for glycine, glutamic and aspartic acids in synaptosomes of rat central nervous system. Brain Res 42: 413–431

Lowenstein O, Osborne MP, Wersäll J (1964) Structure and innervation of the sensory epithelia of the labyrinth in the thornback ray (*Raja clavata*). Proc R Soc Lond B 160: 1–12

Lowenstein O, Osborne MP, Thornhill RA (1968) The anatomy and ultrastructure of the labyrinth of the lamprey (*Lampetra fluviatilis*). Proc R Soc Lond B 170: 113–134

Lu SM, Schweitzer L, Cant NB, Dawbaru D (1987) Immunoreactivity to calcitocin gene-related peptide in the superior olivary complex and cochlea of cat and rat. Hear Res 31: 137–146

Luciano L, Franke K, Iurata S, Reale E (1977) Freeze-fracture study of the cell junctions in the utricle and saccule. Acta Oto-Laryngol 83: 79–84

MacDermott AB, Dale N (1987) Receptors, ion channels and synaptic potentials underlying the integrative actions of excitatory amino acids. Trends Neurosci 10: 280–284

MacDonald JF, Porietis AV, Wojtowicz JM (1982) L-aspartatic acid induced a region of negative slope conductance in the current-voltage relationship of cultured spinal cord neurones. Brain Res 237: 248–253

MacDonald JF, Miljkovic Z, Pennfather P (1987) Use-dependent block of excitatory amino acid currents in cultured neurones by ketamine. J Neurophysiol 58: 251–266

Machin KE (1962) Electric receptors. Abstr Symp Soc Exp Biol, vol 16. Cambridge Univ Press, pp 227–244

Maeno T, Edwards C, Ankaru M (1977) Permeability of the end-plate membrane activated by acetylcholine to some organic cations. J Neurobiol 8: 173–184

Magazanick LG, Antonov SM, Fyodorova IM, Volkova TM, Grishin SV (1986) Effects of the spider Argiope lobata crude venom and its low molecular weight component, argiopin, on the postsynaptic membrane. Biol Membr (USSR) 3: 1204–1219 (in Russian)

Mandelshtam YuYe (1983) Neuron and muscle of insect. Nauka, Leningrad (in Russian)

Martin AR, Ringham GL (1975) Synaptic transfer of a vertebrate central nervous system synapse. J Physiol (Lond) 251: 409–426

Martin AR, Wickelgren WO, Beranek R (1970) Effects of iontophoretically applied drugs on spinal interneurons of the lamprey. J Physiol (Lond) 207: 653–665

Martin D, Lodge D (1985) Ketamine acts as a noncompetitive N-methyl-D-aspartate antagonist on frog spinal cord in vitro. Neuropharmacology 24: 999–1003

Matsuura S, Ikeda K, Furukawa T (1971) Effects of Na^+, K^+ and ouabain on microphonic potentials of the goldfish innear ear. Jpn J Physiol 21: 563–578

Matthews G, Wickelgren WO (1979) Glutamate and synaptic excitation of reticulospinal neurones of lamprey. J Physiol (Lond) 293: 417–433

Mayer ML, Westbrook GL, Guthrie PB (1984) Voltage-dependent block by Mg of NMDA responses in spinal cord neurones. Nature 309: 261–263

McCulloch RM, Johnston GAR, Game CJA, Curtis DR (1974) The differential sensitivity of spinal interneurones and Renshaw cells to kainate and N-methyl-D-aspartate. Exp Brain Res 2: 515–518

McGlone FP, Russell IJ, Sand O (1979) Measurement of calcium ion concentrations in the lateral line cupulae of *Xenopus laevis*. J Exp Biol 83: 123–130

McLennan H (1981) On the nature of the receptors for various excitatory amino acids in the mammalian central nervous system. In: DiChiara G, Gess GL (eds) Glutamate as a neurotransmitter: advances in biochemical psychopharmacology, vol 7. Raven, New York, pp 253–262

McLennan H, Lodge D (1979) The antagonism of amino acid-induced excitation of spinal neurones in the cat. Brain Res 169: 83–90

Medina JE, Drescher DG (1981) The amino-acid content of perilymph and cerebrospinal fluid from guinea pig and the effect of noise on the amino acid composition of perilymph. Neuroscience 6: 505–509

Meister A (1979) Biochemistry of glutamate: glutamine and glutathione. In: Keller LJ Jr, Garattini S, Kane MR, Reynolds WA, Wurtman RJ (eds) Glutamic acid: advances in biochemistry and physiology. Raven, New York, pp 69–84

Melamed B, Morris C, Bryant G, Guth PS (1982) Amino-acid content of guinea pig perilymph collected under conditions of quiet or sound stimulation. Hear Res 7: 13–18

Meza G (1984) Some characteristics of glutamic acid decarboxylase of chick ampullary cristae. J Neurochem 43: 634–639

Meza G (1985) Characterization of GABA-ergic and cholinergic neurotransmission in the chick inner ear. In: Drescher DG (ed) Auditory diochemistry. Charles C Thomas, Springfield, pp 80–101

Meza G, Carabez A, Ruiz M (1982) GABA synthesis in isolated vestibullary tissue of chick inner ear. Brain Res 241: 157–161

Meza G, Gonzales-Viveros T, Ruiz M (1985) [^{3}H] aminobutyric acid binding to vestibular membranes of the chick inner ear. Brain Res 337: 179–183

Michaelis EK, Michaelis ML, Boyarsky LL (1974) High-affinity glutamic acid binding to brain synaptic membranes. Biochem Biophys Acta 367: 338–348

Miledi R, Molenaar PC, Polak RL (1977) An analysis of acetylcholine in frog muscle by mass fragmentography. Proc R Soc Lond B 197: 285–297

Monaghan DT, Bridges RJ, Cotman CW (1989) The excitatory amino acid receptors: their classes, pharmacology, and distinct properties in the function of the central nervous system. Annu Rev Pharmacol Toxicol 29: 365–402

Monaghan P (1975) Ultrastructural and pharmacological studies on the afferent synapse of the lateral line sensory cells of the African clawed toad, *Xenopus laevis*. Cell Tissue Res 163: 239–247

Montgomery J (1984) Noise cancellation in the electrosensory system of the thornback ray: common mode rejection of input produced by the animal's own ventilatory movement. J Comp Physiol 155: 102–112

Mroz EA, Sewell W (1989) Pharmacological alteration of the activity of afferent fibers innervating hair cells. Hear Res 38: 141–162

Mulder AH, Snyder SH (1974) Potassium-induced release of amino acids from cerebral cortex and spinal cord slices of the rat. Brain Res 76: 297–308

Mullinger AM (1964) The fine structure of ampillary electric receptors in *Amiurus*. Proc R Soc Lond B 160: 345–350

Mullinger AM (1969) The organization of ampullary sense organs in the electric fish, *Gymnarchus niloticus*. Tissue Cell 1: 31–52

Mulroy MJ, Altman DW, Weiss TF, Peake WT (1974) Intracellular electric responses to sound in a vertebrate cochlea. Nature 249: 482–485

Munz H, Claas B, Fritzsch B (1982) Electrophysioloigical evidence of electroreception in the axolotl *Siredon mexicanum*. Neurosci Lett 28: 107–111

Murray RW (1959) The response of the ampullae of Lorenzini to combined stimulation to temperature change and weak direct currents. J Physiol (Lond) 145: 1–13

Murray RW (1960) The response of the ampullae of Lorenzini of elasmobranchs to mechanical stimulation. J Exp Biol 37: 417–424
Murray RW (1962a) The response of the ampullae of Lorenzini of elasmobranchs to electrical stimulation. J Exp Biol 39: 119–128
Murray RW (1962b) Temperature receptors in animals. Abstr Symp Soc Exp Biol, vol 16. Cambridge Univ Press, pp 245–266
Murray RW Receptor mechanisms in the ampullae of Lorenzini of the elasmobranch fishes. Cold Spring Harbor Symp Quant Biol 30: 233–243
Murray RW (1974) The ampullae of Lorenzini. In: Fessard A (ed) Electroreceptors and other specialized receptors in lower vertebrates. Springer, Berlin Heidelberg New York, pp 125–146 (Handbook of sensory physiology, vol III/3)
Murray RW, Potts WTW (1961) The composition of the endolymph perilymph and other body fluids of elasmobranchs. Comp Biochem Physiol 2: 65–75
Nadler JV (1979) Kainic acid: neurophysiological and neurotoxic actions. Life Sci 24: 289–300
Nadol JB (1981) Reciprocal synapses at the base of outer hair cells in the organ of Corti. Ann Otol Rhinol Laryngol 90: 12–17
Nagai T, Obara S (1985) Afferent facilitation induced by iontophoretic application of acidic amino acids in the ampullary electroreceptors of *Plotosus*. J Neurophysiol 53: 853–867
Nagai T, Obara S, Kawai N (1984) Differential blocking effects of a spider toxin on synaptic and glutamate responses in the afferent synapse of the acoustico-lateralis receptors of *Plotosus*. Brain Res 300: 183–187
Naito S, Ueda T (1983) ATP-dependent uptake of glutamate into protein I-associated synaptic vesicles. J Biol Chem 258: 696–699
Naito S, Ueda T (1985) Characterization of glutamate uptake into synaptic vesicles. J Neurochem 44: 99–109
Nakagawa T, Kakehata S, Akaike N, Komune S, Takasaka T, Uemura T (1991) Calcium channel in isolated outer hair cells of guinea pig cochlea. Neuroci Lett 125: 81–84
Neal MJ (1971) The uptake of [^{14}C]glycine by slices of mammalian spinal cord. J Physiol (Lond) 215: 103–117
Nelson PG, Pun RYK, Westbrook GL (1986) Synaptic excitation in cultures of mouse spinal cord neurones: receptor pharmacology and behaviour of synaptic currents. J Physiol (Lond) 372: 169–190
Nicholls DG, Sihra TS, Sanchez-Prieto J (1987) Calcium-dependent and -independent release of glutamate from synaptosomes monitored by continuous fluorometry. J Neurochem 49: 50–57
Nier K, Hensel H (1973) Effects of calcium on afferent responses of the ampullae of Lorenzini. Pfluegers Arch 338: 281–287
Nier K, Hensel H, Bromm B (1976) Differential thermosensitivity and electric polarization of the ampullae of Lorenzini. Pfluegers Arch 363: 181–185
Norris CH, Guth PS (1974) The release of acetylcholine (ACh) by the crossed olivocochlear (COCS) bundle. Acta Oto-Laryngol 77: 318–326
Norris CH, Guth PS, Hastings RC (1976) The infusion of picrotoxin, bicuculine, amino-oxyacetic acid, and several putative transmitters on afferent transmission. Fed Proc 35: 621
Norris CH, Housley GD, Williams WH, Guth SL, Guth PS (1988) The acetylcholine receptors of the semicircular canal in the frog (*Rana pipiens*). Hear Res 32: 197–206
Norris CH, Ricci AJ, Housley GD, Guth PS (1992) The inactivating potassium currents of hair cells isolated from the crista ampullaris of the frog. J Neurophysiol 68: 1642–1653
Nowak L, Bregestovski P, Ascher P, Herbert A, Prochiantz A (1984) Magnesium gates glutamate-activated channels in mouse central neurons. Nature 307: 462–465
Obara S (1976) Mechanism of electroreception in ampullae of Lorenzini of the marine catfish *Plotosus*. In: Reuben JP, Purpura DP, Bennett MVL, Lendall ER (eds) Electrobiology of nerve, synapse and muscle. Raven, New York, pp 129–147

Obara S, Bennett MVL (1972) Mode of operation of ampullae of Lorenzini of the skate, *Raja*. J Gen Physiol 60: 534–557
Obara S, Oomura Y (1973) Disfacilitation as the basis for sensory suppression in a specialized lateralis receptor of the marine catfish. Proc Jpn Acad 49: 213–217
Obara S, Sugawara Y (1979) Contribution of Ca^{2+} to the electroreceptor mechanism in *Plotosus* ampullae. J Physiol (Paris) 75: 335–340
Obara S, Sugawara Y (1984) Electroreceptor mechanisms in teleost and non-teleost fishes. In: Bolis L, Keynes RD, Maddrell SHP (eds) Comparative physiology of sensory systems. Combridge Univ Press, pp 509–523
O'Brien RJ, Fischbach GD (1986) Characterization of excitatory amino acid receptors expressed by embryonic chick motoneurones in vitro. J Neurosci 6: 3275–3283
Ohmori H (1984a) Studies of ionic currents in the isolated vestibular hair cell of the chick. J Physiol (Lond) 350: 561–581
Ohmori H (1984b) Mechano-electrical transducer has discrete conductances in the chick vestibular hair cell. Proc Natl Acad Sci USA 81: 1888–1891
Ohmori H (1985) Mechano-electrical transduction currents in isolated vestibular hair cell of the chick. J Physiol (Lond) 359: 189–217
Ohmori H (1987) Gating properties of the mechano-electrical transducer channel in the dissociated vestibular hair cell of the chick. J Physiol (Lond) 387: 589–609
Ohmori H (1988a) Mechano-electric transduction of the chick hair cell. Prog Brain Res 74: 11–20
Ohmori H (1988b) Mechanical stimulation and Fura-2 fluorescence in the hair bundle of dissociated hair cells of the chick. J Physiol (Lond) 399: 115–137
Ohmori H (1989) Mechano-electrical transduction of the hair cell. Jpn J Physiol 39: 643–657
Okano K (1988) Two pharmacologically distinct receptors mediate two postsynaptic potentials (PSP) components in the afferent synapse of the *Plotosus* electroreceptor. Brain Res 457: 89–97
Okano K, Obara S (1988) Effect of bath-applied L-glutamate and related chemicals on the afferent synapse of the *Plotosus* electroreceptor. Brain Res 457: 79–88
Okano K, Sugawara Y, Obara S (1984) Suppressive effect of S-HT on the ampullary electroreceptor of *Plotosus*. Jpn J Physiol 46: 448
Okitsu S, Umekita S, Obara S (1978) Ionic compositions of the media across the sensory epithelium in the ampullae of Lorenzini of the marine catfish *Plotosus*. J Comp Physiol 126: 115–121
Onodera K, Takeuchi A (1975) Ionic mechanism of the excitatory synaptic membrane of the crayfish neuromuscular junction. J Physiol (Lond) 252: 295–318
Orman S, Flock Å (1983) Active control of sensory hair mechanics implied by susceptibility to media that induce contraction in muscle. Hear Res 11: 261–266
Orrego F (1979) Criteria for the identification of central neurotransmitters and their application to studies with nerve tissue preparations in vitro. Neuroscience 4: 1037–1057
Orrego F, Miranda R, Saldate C (1976) Electrically induced release of labelled taurine, - and - alanine, glycine, glutamate and other amino acids from rat neocortical slices in vitro. Neuroscience 1: 325–332
Osborne MP (1977) Role of vesicles with some observations on vertebrate sensory cells. In: Gottrell GA, Usherwood PNP (eds) Synapses. Academic Press, New York, pp 40–63
Osborne MP, Thornhill RA (1972) The effect of monoamine depleting drugs upon the synaptic hairs in the inner ear of the bullfrog (*Rana catesbeiana*). Z Zellforsch Mikrosk Anat 127: 347–355
Pappas GD (1975) Ultrastractural basis of synaptic transmission. In: Tower DB, Brady RO (eds) The nervous system. The basic neurosciences, vol 1. Raven, New York, pp 19–30
Pastuszko A, Wilson DP, Erecinska M (1984) Effect of kainic acid in rat brain synaptosomes: the involvement of calcium. J Neurochem 43: 747–754

Pennfather P, Quastel MJ (1982) Modification of dose-response curves by effector blockade and uncompetitive antagonism. Mol Pharmacol 22: 369–380

Pèrez MA, Soto E, Vega R (1991) Streptomycin blocks the postsynaptic effects of excitatory amino acids on the vestibular system primary afferents. Brain Res 563: 221–226

Perkins MN, Stone TW (1982) An iontophoresis investigation of the actions of convulsant kynurenines and the interaction with the endogenous excitant quinolinic acid. Brain Res 247: 184–187

Perney TM, Dinerstein RJ, Miller RJ (1984) Depolarization-induced increases in intracellular free calcium detected in single cultured neuronal cells. Neurosci Lett 51: 165–170

Peters RC, Buwalda RJA (1972) Frequency response of the electroreceptors "small pit organs" of the catfish, *Ictalurus nebulosus* LeS. J Comp Physiol 79: 29–38

Peters RC, Loos WJG, Gerritsen A (1974) Distribution of electroreceptors, bioelectric field patterns, and skin resistance in the catfish, *Ictalurus nebulosus* LeS. J Comp Physiol 92: 11–22

Peters RC, Bretschneider F, Schreuder J-JA (1975) Influence of direct current stimulation on the ion-induced sensitivity changes of the electroreceptors (small pit organs) of the brown bullhead, *Ictalurus nebulosus* LeS. Neth J Zool 25: 389–397

Peters RC, Groen EL, van der Sluis MM, Teunis PFM, Wilhelm KA (1988) Denervation reveals two components of neurotransmission in electroreceptor synapses. Neuroscience 27: 1049–1053

Peterson NA, Raghupathy E (1978) Evidence for strictly Na^+-dependent accumulation of serine and threonine in brain cortical synaptosomes from newborn rats. J Neurochem 31: 989–996

Pickles JO, Comis SD, Osborne MP (1984) Cross-links between stereocilia in the guinea pig organ of Corti, and their possible relation to sensory transduction. Hear Res 15: 103–112

Pickles JO, Corey DP (1992) Mechanoelectrical transduction by hair cells. Trends Neuroscience 15: 254–259

Pierau F-K, Meissner HDS (1966) A contribution to the temperature sensitivity of Lorenzinian ampullae of elesmobranchs. Experientia 22: 1–4

Pierau F-K, Torrey P, Carpenter DO (1974) Mammalian cold receptor afferents: role of an electrogenic sodium pump in sensory transduction. Brain Res 73: 156–160

Piotrovsky LB (1987) Excitatory amino acids and their antagonists (structure and activity). Chem-Pharm J USSR 7: 773–782 (in Russian)

Plinkert PK, Gitter AH, Zimmerman U, Kirschner T, Tzartos S, Zenner HP (1990) Visualisation and functional testing of acetylcholine receptor-like molecules in cochlear hair cells. Hear Res 44: 25–34

Prigioni I, Russo G, Valli P, Masetto S (1990) Pre- and postsynaptic excitatory action of glutamate agonists on frog vestibular receptors. Hear Res 46: 253–260

Prigioni I, Masetto S, Russo G, Taglietti V (1992) Calcium currents in solitary hair cells isolated from frog crista ampullaris. J Vestib Res 2: 31–39

Puel J-L, Bobbin RP, Fallon M (1988) 2-Amino-4-phosphonobutiric acid receptors are not involved in synaptic transmission from hair cells to auditory neurons. Hear Res 37: 83–88

Puel J-L, Bobbin RP, Fallon M (1989) Suppression of auditory nerve activity in the guinea pig cochlea by 1-(p-bromobenzoyl)-piperazine-2,3-dicarboxylic acid. Brain Res 487: 9–15

Puel J-L, Ladrech S, Chabert R, Pujol R, Eybalin M (1991) Electrophysiological evidence for the presence of NMDA receptors in the guinea pig cochlea. Hear Res 51: 255–264

Puel J-L, Pujol R, Ladrech S, Eybalin M (1991)a-Amino-3-hydroxy-5-methyl-4-isoxazole propionic acid electrophysiological and neurotoxic effects in the guinea-pig cochlea. Neuroscience 45: 63–72

Pujol R, Lenoir M (1986) The four types of synapses in the organ of Corti. In: Altschuler RA, Hoffman DW, Bobbin RP (eds) Neurobiology of hearing: the cochlea. Raven, New York, pp 161–172

Pujol R, Lenoir M, Robertson D, Eybalin M, Johnstone BM (1985) Kainic acid selectively alters auditory dendrites connected with cochlear inner hair cells. Hear Res 18: 145–152
Ransom BR, Bullock PN, Nelson PG (1977) Mouse spinal cord in cell culture. III. Neuronal chemosensitivity and its relationship to synaptic activity. J Neurophysiol 40: 1163–1177
Raschi W (1986) A morphological analysis of the ampullae of Lorenzini in selected skates (Pisces, Rajoidei). J Morphol 189: 225–247
Raschi W, Mackanos LA (1987) Evolutionary trends in the peripheral component of chondrichthyian electroreceptors: a morphological review. Arch Biol 98: 163–186
Raymond J, Nieoullon A, Dememes D, Sans A (1984) Evidence for glutamate as a neurotransmitter in the cat vestibular nerve:
Rebillard G, Bryant GM (1989) Effects of in vivo perfusion of glutamate dehidrogenase in the guinea pig cochlea on the VIII the nerve compound action potential. Brain Res 494: 379–382
Redburn DA, Shelton D, Cotman CW (1976) Calcium-dependent release of exogenously loaded -amino- U- C butyrate from synaptosomes: time course of stimulation by potassium, veratridine, and the calcium ionophore A 23187. J Neurochem 26: 297–303
Rennie KJ, Ashmore JF (1991) Ionic currents in isolated hair cells from the guinea-pig crista ampullaris. Hear Res 51: 279–292
Reubi JC (1980) Comparative study of the release of glutamate and GABA, newly senthesized from glutamine, in various regions of CNS. Neuroscience 5: 2145–2150
Reuter G, Zenner H-P (1990) Active radial and transverse motile responses of outer hair cells in the organ of Corti. Hear Res 43: 219–230
Roberts PJ (1974) Glutamate receptors in the rat central nervous system. Nature 252: 399–401
Roberts PJ (1981) Binding studies for the investigation of receptors for L-glutamate and other excitatory amino acids. In: Roberts PJ, Storm-Mathiesen J, Johnston GAR (eds) Glutamate: transmitter in the central nervous system. Wiley, Chichester, pp, 35–53
Roberts PJ (1984) Receptors for excitatory amino acids and excitotoxins. In: Melchiorre C, Gianelle M (eds) Highlights in receptor chemistry. Elsevier, Amsterdam, pp 169–184
Roberts BL, Ryan K (1971) The fine structure of the lateral line sense organs of dogfish. Proc R Soc Lond B 179: 157–168
Roberts WM, Howard J, Hudspeth AJ (1988) Mechanoelectrical transduction, frequency tuning and synaptic transmission by hair cells. Annŭ Rev Biol 4: 63–92
Roberts WM, Jacobs RA, Hudspeth AJ (1990) Colocalization of ion channels involved in frequency selectivity and synaptic transmission at presynaptic active zones of hair cells. J Neurosci 10: 3664–3684
Roberts WM, Jacobs RA, Hudspeth AJ (1991) The hair cell as a presynaptic terminal. Ann N Y Acad Sci 635: 221–233
Robertson D, Johnstone BM (1978) Efferent transmitter substance in the mammalian cochlea: single neuron support for acetylcholine. Hear Res 1: 31–34
Robertson D, Johnstone BM (1979) Effects of divalent cations on spontaneous and evoked activity of single mammalian auditory neurons. Pfluegers Arch 380: 7–12
Rossi ML, Martini M (1988) High perilymphatic potassium and rubidium levels modulate the sensory resting discharge in the isolated frog labyrinth. Neurosci Lett 93: 44–48
Rossi ML, Valli P, Casella C (1977) Postsynaptic potentials recorded from afferent nerve fibres of the posterior semicircular canal in the frog. Brain Res 135: 67–75
Rossi ML, Prigioni J, Valli P, Casella C (1980) Activation of the efferent system in the isolated frog labyrinth: effects on the afferent EPSPs and spike discharge recorded from single fibres of the posterior nerve. Brain Res 185: 125–137
Rossi ML, Bonifazzi C, Martini M, Fesse R (1989) Static and dynamic properties of synaptic transmission at the cyto-neural junction of frog posterior canal. J Gen Physiol 94: 303–327

Roth A (1971) Zur Funktionweise der Electrorezeptoren in der Haut von Welsen (*Ictalurus*): der Einfluss der Ionen im Süsswasser. Z Vgl Physiol 75: 303–322
Roth A, Szabo T (1969) The effect of sensory nerve transsection on the sensory cells and on the receptor potential of the tuberous (Knollen) organ in mormyrid fish (*Gnathonemus* sp.). Z Vgl Physiol 62: 395–410
Roth A, Szabo T (1970) Qualitative analysis of the effect thermal stimuli on the spontaneous activity of tuberous (Knollen) organs in *Gnathonemus* sp., Mormyridae, Pisces. J Exp Biol 52: 707–719
Roth A, Tscharntke H (1976) Ultrastructure of the ampullary electroreceptors in lungfish and Branchiopterigii. Cell Tissue Res 173: 95–108
Runhaar G, Manley GA (1987) Potassium concentration in the inner sulcus is perilymphlike. Hear Res 29: 93–103
Rüsch A, Thurm U (1990) Spontaneous and electrically induced movements of ampullary kinocilia and stereovilli. Hear Res 48: 247–264
Russell IJ (1968) Influence of efferent fibres on a receptor. Nature 219: 177–178
Russell IJ (1971) The pharmacology of efferent synapses in the lateral-line system of *Xenopus laevis*. J Exp Biol 54: 643–658
Russell IJ (1976) Amphibian lateral line receptors. In: Llinas R, Precht W (eds) Frog neurobiology. Springer, Berlin Heidelberg New York, pp 543–549
Russell IJ (1983) Origin of the receptor potential in inner hair cells of the mammalian cochlea-evidence for Davis' theory. Nature 301: 334–336
Russell IJ, Richardson GP (1987) The morphology and physiology of hair cells in organotypic cultures of the mouse cochlea. Hear Res 31: 9–24
Russel IJ, Sellick PM (1976) Measurement of potassium and chloride ion concentration in the cupula of the lateral lines of *Xenopus laevis*. J Physiol (Lond) 257: 245–255
Russell IJ, Sellick PM (1978) Intracellular studies of hair cells in the mammalian cochlear. J Physiol (Lond) 284: 281–290
Russell IJ, Sellick PM (1981) The response of hair cells to low frequency tones and their relationship to the extracellular receptor potentials and sound pressure level in the guinea pig cochlea. In: Syka J, Aitkin L (eds) Mechanism of hearing. Plenum, New York, pp 3–15
Russell IJ, Sellick PM (1983) Low-frequency characteristics of intracellular recorded receptor potentials in guinea pig cochlea hair cells. J Physiol (Lond) 338: 179–206
Russell IJ, Richardson GP, Cody AR (1986a) Mechanosensitivity of mammalian auditory hair cells in vitro. Nature 321: 517–519
Russell IJ, Cody AR, Richardson GP (1986b) The responses of inner and outer hair cells in the basal turn of the guinea-pig cochlea and in the mouse cochlea grown in vitro. Hear Res 22: 199–216
Russell IJ, Kössl M (1991) The voltage respouses of hair cells in the basal turn of the guinea-pig cochlea. J Physiol 435: 493–511
Ryan AJ, Schwartz JR (1984) Preferential glutamate uptake by cochlear hair cells: implications for the afferent cochlear transmitter. Brain Res 290: 376–379
Saito K (1980) Fine structure of the sensory epithelium of the guinea pig organ of Corti: afferent and efferent synapses of hair cells. J Ultrastruct Res 71: 222–232
Sanchez-Prieto J, Sihra TS, Nicholls DG (1987) Characterization of the exocytotic release of glutamate from guinea-pig cerebral cortical synaptosomes. J Neurochem 49: 58–64
Sand A (1938) The function of the ampullae of Lorenzini, with some observation on the effect of temperature on sensory rhythms. Proc R Soc Lond B 125: 524–553
Sand O (1975) Effects of different ionic environment on the mechanosensitivity of lateral line organs in the mudpuppy. J Comp Physiol 102: 27–42
Sando I (1965) The anatomical interrelationships of the cochlear nerve fibers. Acta Oto-Laryngol 59: 417–428
Santi PA, Andersson CB (1987) A newly identified surface coat on cochlear hair cells. Hear Res 27: 47–65
Santos-Sacchi J (1989) Asymmetry in voltage-dependent movements of isolated outer hair cells from the organ of Corti. J Neurosci 9: 2954–2962

Santos-Sacchi J, Dilger JP (1988) Whole cell currents and mechanical responses of isolated outer hair cells. Hear Res 35: 143–150
Sariban-Sohraby S, Benos D (1986) The amiloride-sensitive sodium channel. Am J Physiol 250: 175–190
Scarfone E, Ulfendahl M, Löfstrand P, Flock Å (1991) Light- and electron microscopy of isolated vestibular hair cells from the guinea pig. Cell Tissue Res 266: 51–58
Scatchard G (1949) The attractions of proteins for small molecules and ions. Ann NY Acad Sci 51: 660–672
Schacht J, Zenner H-P (1987) Evidence that phosphoinositides mediate motility in cochlear outer hair cells. Hear Res 31: 155–160
Schank RP, Schneider CR, Tighe JJ (1987) Ion dependence of neurotransmitter uptake: inhibitory effect of ion substitutes. J Neurochem 49: 381–388
Scheich H, Langer G, Tideman C, Coles RB, Cuppy A (1986) Electroreception and electrolocation in platypus. Nature 319: 401–402
Schessel DA, Ginzburg RD, Highstein SM (1982) Ultrastructural and physiological evidence for chemical synaptic transmission between the vestibular type I hair cell and its primary nerve chalice. Anat Rec 202: 168
Schmahlhausen II (1938) Principles of comparative anatomy of vertebrates. Uchpedgiz, Moscow (in Russian)
Schwartz AM (1986) Auditory nerve and spiral ganglion cells: morphology and organization. In: Altschuler RA, Hoffman DW, Bobbin RP (eds) Neurobiology of hearing: the cochlea. Raven, New York, pp 271–282
Schwartz IR, Ryan AF (1983) Differential labelling of sensory cells and neuronal populations in the organ of Corti following amino acid incubation. Hear Res 9: 185–200
Sejnowski TJ, Yodlowski ML (1982) A freeze-fracture study of the skate electroreceptor. J Neurocytol 11: 897–912
Sellick PM, Johnstone BM (1975) Production and role of inner ear fluid. Prog Neurobiol 5: 337–362
Sewell WF (1990) Synaptic potentials in afferent fibers innervating hair cells of the lateral line organ in *Xenopus laevis*. Hear Res 44: 71–82
Sewell WF, Mroz EA (1987a) Neuroactive substances in inner ear extracts. J Neurosci 7: 2465–2475
Sewell WF, Mroz EA (1987b) Isolation of neuroactive substances from hair cell extracts. Soc Neurosci Abstr 13: 42
Sewell WF, Norris CH, Tachibana M, Guth PS (1978) Detection of an auditory nerve-activating substance. Science 202: 910–912
Sewell WF, Starr PA (1991) Effects of calcitonin gene-related peptide and efferent nerve stimulation on afferent transmission in the lateral line organ. J Neurophysiol 65: 1158–1169
Shapovalov AI (1966) Cellular mechanisms of synaptic transmission. Nauka, Moscow (in Russian)
Shapovalov AI, Shiriayev BI, Velumian AA (1978) Mechanisms of postsynaptic excitation in amphibian motoneurons. J Physiol (Lond) 279: 437–455
Shelton PMJ (1971) The structure and function of the lateral line system in larval *Xenopus laevis*. J Exp Zool 178: 211–218
Shigemoto T, Ohmori H (1990) Muscarinic agonists and ATP increase the intracellular Ca^{2+} coccentration in chick cochlear hair cells. J Physiol (Lond) 420: 127–148
Shinozaki H (1988) Pharmacology of the glutamate receptor. Prog Neurobiol 30: 399–435
Siegel JU, Relkin EM (1987) Antagonistic effects of perilymphatic calcium and magnesium on the activity of single cochlear afferent neurous. Hear Res 28: 131–147
Sliwinska-Kowalska M, Parrakkal M, Schneider ME, Fex J (1989) CGRP-like immunoreactivity in the guinea pig organ of Corti: a light and electron microscopy study. Hear Res 42: 83–96
Smith CA (1956) Microscopic structure of the utricle. Ann Otol (St Louis) 65: 450–469
Smith CA (1961) Innervation pattern of the cochlea. The internal hair cell. Ann Otol (St Louis) 70: 1–24

Smith CA (1967) Utricle and saccule. In: Iurato S (ed) Submicroscopic structure of the inner ear. Pergamon, Oxford, pp 107–131

Smith CA, Dempsey EW (1957) Electron microscopy of the organ of Corti. Am J Anat 100: 337–368

Smith CA, Rasmussen GL (1963) Recent observations on the olivocochlear bundle. Ann Otol (St Louis) 78: 489–506

Smith CA, Rasmussen GL (1965) Degeneration in the efferent nerve endings in the cochlea after axonal section. J Cell Biol 26: 63–77

Smith CA, Sjöstrand FS (1961) Structure of nerve endings on the external hair cells of the guinea pig cochlea by serial sections. J Ultrastruct Res 5: 523–556

Sokolove PG, Cooke JM (1971) Inhibition of impulse activity in sensory neurons by an electrogenic pump. J Gen Physiol 57: 125–163

Sonnhof U, Buhrle ChPh (1980) On the postsynaptic action of glutamate in frog spinal motoneurons. Pfluegers Arch 388: 101–109

Sonnhof U, Grafe P, Richter DW, Parekh N, Krumnike G, Hinder M (1978) Investigations of the effects of glutamate on mototoneurons of the isolated frog spinal cord. In: Rvall RW, Kelly JS (eds) Iontophoresis in the mammalian central nervous system. Elsevier, Amsterdam, pp 391–393

Soto E, Vega R (1988) Actions of excitatory amino acid agonists and antagonists on the primary afferents of the vestibular system of the axolotl (*Amblystoma mexicanum*). Brain Res 462: 104–111

Soto E, Perez ME, Vega R (1988) Naloxone induces an excitatory effect on the vestibular system primary afferents. Neurosci Soc Abstr 14: 331

Spoendlin H (1964) Organization of the sensory hairs in the gravity receptors in utricle and saccule of the squirrel monkey. Z Zellforsch 62: 701–716

Spoendlin H (1966) The organization of the cochlear receptor. Karger, Basel, 264 pp

Spoendlin H (1969) Innervation patterns in the organ of Corti of the cat. Acta Oto-Laryngol 67: 239–254

Spoendlin H (1972) Innervation densities of the cochlea. Acta Ota-Laryngol 73: 235–248

Spoendlin H (1985) Anatomy of cochlea innervation. Am J Oto-Laringol 6: 453–467

Spoendlin H, Gasek RP (1963) Electronmicroscopic study of the efferent and afferent innervation of the organ of Corti in the cat. Ann Otol (St Louis) 72: 660–686

Spoendlin H, Lichtensteiger W (1966) The adrenergic innervation of the labyrinth. Acta Oto-Laryngol 61: 423–424

Spray DC (1974) Metabolic dependence of frog cold receptor sensitivity. Brain Res 72: 354–359

Srivastava CBL (1972) Morphological evidence for electrical synapse of "gap" junction type in another vertebrate receptor. Experientia 28: 1029–1030

Starr PA, Sewell WF (1991) Neurotransmitter release from hair cells and its blockade by glutamate-receptor antagonists. Hear Res 52: 23–42

Steinacker A, Romero A (1991) Characterization of voltage-gated and calcium-activated potassium currents in toadfish saccular hair cells. Brain Res 556: 22–32

Steinbach AB (1974) Transmission from receptor cells in afferent nerve fibres. In: Bennett MVL (ed) Synaptic transmission and interneuronal communication. Raven, New York, pp 105–140

Steinbach AB, Bennett MVL (1971) Effect of divalent ions and drugs on synaptic transmission in phasic electroreceptors in mormyrid fish. J Gen Physiol 58: 580–598

Steinbach HWM, Verhofstad AAJ, Joosien HWJ (1978) Localization of serotonin in the central nervous system by immunohistochemistry: description of a specific and sensitive technique and some applications. Neuroscience 3: 811–819

Storm-Mathiesen J (1977) Glutamic acid and excitatory nerve endings: reduction of glutamic acid uptake after axotomy. Brain Res 120: 379–386

Storm-Mathiesen J (1981) Autoradiographic and microchemical localization of high affinity glutamate uptake. In: Roberts PJ, Storm-Mathiesen J, Johnston GAR (eds) Glutamate: transmitter in the central nervous system. Wiley, Chichester, pp 89–115

Storm-Mathiesen J, Iversen LL (1979) Uptake of [H]-glutamic acid in excitatory nerve endings. Light and electromicroscopic observation in the hippocampal formation of the rat. Neuroscience 4: 1237–1253

Storm-Mathiesen J, Leknes AK, Bore AT, Vaaland JL, Edminson P, Hang F-MS, Ottersen OP (1983) First visualization of glutamate and GABA in neurons by immunocytochemistry. Nature 301: 517–519

Streit P (1980) Selective retrograde labelling indicating the transmitter of neuronal pathways. J Comp Neurol 191: 429–463

Sugawara Y (1989a) Two Ca^{2+} current components of the receptor current in the electroreceptors of the marine catfish *Plotosus*. J Gen Physiol 93: 365–380

Sugawara Y (1989b) Electrogenic Na^+ $-$-K^+ pump at the basal face of the sensory epithelium in the *Plotosus* electroreceptor. J Comp Physiol 164 (A): 589–596

Sugawara Y Obara S (1984a) Damped oscillation in the ampullary electroreceptors of *Plotosus* involves Ca^{2+}-activated transient K^+ conductance in the basal membrane of receptor cells. Brain Res 302: 171–175

Sugawara Y, Obara S (1984b) Ionic currents in the sensory epithelium examined in isolated electroreceptors of *Plotosus* under simultaneous in situ conditions. Brain Res 302: 176–179

Sugawara Y, Obara S (1989) Receptor Ca^{2+}-current and Ca^{2+}-gated K^+-current in tonic electroreceptors of the marine catfish *Plotosus*. J Gen Physiol 93: 343–364

Sugihara I, Furukawa T (1989) Strontium can activate the Ca^{2+}-dependent K^+ channel in goldfish hair cell membrane. Proc IUPS 16: 420–422

Sugihara I, Furukawa T (1990) Morphological and functional aspects of two different types of hair cells in the goldfish sacculus. J Neurophysiol 62: 1330–1343

Sunose H, Ikeda K, Saito Y, Nishiyama A, Takasaka T (1992) Membrane potential measurement in isolated outer hair cells of the guinea pig cochlea using conventional microelectrodes. Hear Res 62: 237–244

Szabo T (1965) Sense organs of the lateral line system in some electric fish of the Gymnotidae, Mormyridae and Gymnarchidae. J Morphol 117: 229–249

Szabo T (1972) Ultrastructural evidence for mechanoreceptor function of the ampullae of Lorenzini. J Microsc (Paris) 14: 343–350

Szabo T (1974) Anatomy of specialized lateral line organs of electroreception. In: Fessard A (ed) Electroreceptors and other specialized receptors in lower vertebrates. Springer, Berlin Heidelberg New York, pp 13–59 (Handbook of sensory physiology, vol III/3)

Szabo T, Wersäll J (1970) Ultrastructure of an electroreceptor (Mormyromast) in a mormyrid fish *Gnathonemus petersii*. J Ultrastruct Res 30: 473–490

Szabo T, Kalmijn AJ, Enger PS, Bullock TH (1972) Microampullary canals and submandibular sense organs in the fresh-water ray, *Potamotrygon*. J Comp Physiol 79: 15–27

Szamier RB, Bennett MVL (1980) Ampullary electroreceptors in the fresh water ray, *Potamotrygon*. J Comp Physiol 138: 225–230

Szamier RB, Wachtal AW (1969) Special cutaneous receptor organs of fish. III. Ampullary organs of *Eigenmannia*. J Morphol 119: 261–290

Szamier RB, Wachtal AW (1970) Special cutaneous receptor organs of fish. VI. Ampullary and tuberous organs of *Hypopomus*. J Ultrastruct Res 30: 450–471

Tachibana M, Kuriyama K (1974) Gamma-aminobutyric acid in the lower auditory pathway of the guinea pig. Brain Res 69: 370–374

Taglietti V, Rossi ML, Casella C (1977) Adaptive distortions in the generator potential of semicircular canal sensory afferent. Brain Res 123: 41–57

Takeda N, Kitajiri M, Girgis S, Hillyard CJ, MacIntyre I, Emson PC, Shinosaka S, Toyama M, Matsunaga T (1986) The presence of a calcitonin gene-related peptide in the olivocochlear bundle in rat. Exp Brain Res 61: 575–578

Takeda N, Doi K, Mori N, Yamazaki H, Toyama M, Matsunaga T (1987) Localization and fine structure of calcitonin gene-related peptide (CGRP)-like immunoreactive nerve fibres in the organ of Corti of guinea-pigs by immunohistochemistry. Acta Oto-Laryngol 103: 567–571

Takeuchi A (1987) The transmitter role of glutamate in nervous system. Jpn J Physiol 37: 559–572

Takeuchi A, Takeuchi N (1962) Electrical changes in pre- and postsynaptic axons of the giant synapse of *Loligo*. J Gen Physiol 45: 1181–1191

Tanaka Y, Asanuma A, Yanigisawa K (1980) Potentials of outer hair cells and their membrane properties in cationic environments. Hear Res 2: 431–438

Teeter JH, Bennett MVL (1981) Synaptic transmission in the ampullary electroreceptors of the transparent catfish, *Kryptopterus*. J Comp Physiol (A) 142: 371–377

Teeter JH, Szamier RB, Bennett MVL (1980) Ampullary electroreceptors in the sturgeon *Scaphirhynchus platorynchus* (Rafinesque). J Comp Physiol 138: 213–223

Tester AK, Kendall JJ (1969) Morphology of the lateral line canal system in the shark genus *Carchinus*. Pac Sci 23: 1–16

Thornhill RA (1972) The effect of catecholamine precursors and related drugs on the morphology of the synaptic bars in the vestibular epithelia of the frog, *Rana temporaria*. Comp Gen Pharmacol 3: 89–97

Thompson GC, Cortez AM, Igarashi M (1986) GABA-like immunoreactivity in the squirrel monkey organ of Corti. Brain Res 372: 72–79

Thomson AM (1986) A magnesium-sensitive post-synaptic potential in rat cerebral cortex resembles neuronal responses to N-methylaspartate. J Physiol (Lond) 370: 531–549

Toyoshima K, Shimamura A (1982) Comparative study of ultrastructures of the lateral-line organs in the African clawed toad, *Xenopus laevis*. Anat Rec 204: 371–381

Umekita S-I, Matsumoto Y, Abe T, Obara S (1980) The afferent neurotransmitter in the ampullary electroreceptors: stimulus-dependent release experiments refute the transmitter role of L-glutamate. Neurosci Lett Suppl 4: S7

Usami S, Igarashi M, Thompson GC (1987) GABA-like immunoreactivity in the chick vestibular end organs. Brain Res: 383–387

Usami S, Igarashi M, Thompson GC (1988) Light and electron microscopic study of gamma aminobutyric acid–like immunoreactivity in the guinea-pig organ of Corti. J Otorhinolaryngol 50: 162–169

Usherwood PNR, Machili P, Leaf G (1968) L-glutamate at insect excitatory nerve-muscle synapses. Nature 219: 1169–1172

Valli P, Taglietti V, Rossi ML (1974) Effects of d-tubocurarine on the ampullar receptors of the frog. Acta Oto-Laryngol 78: 51–58

Valli P, Zucca G, Casella C (1977) The importance of potassium in the function of frog semicircular canals. Acta Oto-Laryngol 84: 344–351

Valli P, Zucca G, Casella C (1979) Ionic composition of the endolymph and sensory transduction in labyrinthine organs. Acta Oto-Laryngol 87: 466–471

Valli P, Zucca G, Botta L, Casella C (1984) Specificity of the K^+-channels in labyrinthine organs: effects of Rb^{2+} and Ca^{2+} on the conversion process in frog semicircular canals. J Comp Physiol (A) 155: 739–743

Valli P, Zucca G, Prigioni I, Botta L, Gasella C, Guth P (1985) The effect of glutamate on the frog semicircular canal. Brain Res 330: 1–9

Valli P, Zucca G, Botta L, Casella C (1988) Sensitivity and adaptation of ampullar receptors to rapid perilymphatic potassium changes. J Comp Physiol (A) 162: 173–178

Valli P, Zucca G, Botta L (1990) Perilymphatic potassium changes and potassium homeostasis in isolated semicircular canals of the frog. J Physiol (Lond) 430: 585–594

Vega R, Soto E, Budelli R, Gonzalez-Estrada ML (1987) Is GABA an afferent transmitter in the vestibular system? Hear Res 29: 163–168

Vinnikov TA (1979) Peripheral electrosense physiology: a review of recent findings. J Physiol (Paris) 75: 321–333

Vinnikov YaA (1971) Cytologic and molecular foundations of reception. Nauka, Leningrad (in Russian)

Vinnikov YaA (1979) Evolution of receptors. Nauka, Leningrad (in Russian)

Vinnikov YaA, Titova LK (1961) The organ of Corti (histophysiology and histochemistry). Nauka, Leningrad (in Russian)

Vinnikov YaA, Govardovsky VI, Osipova IV (1965) Substructural organization of the gravitation organ – pigeon utricules. Biophysics (USSR) 10: 641–644 (in Russian)

Vinnikov YaA, Gazenko OG, Titova LK, Bronstein AA, Tsirulis TP, Pevsner RA, Govardovsky BU, Gribakin FG, Ivanov VP, Aronova MZ, Thechonadsky NA (1971) The gravity receptor. Evolution of structure, cytochemical and functional organization. Nauka, Leningrad (in Russian)
Wachtel AW, Szamier RB (1966) Special cutaneous receptor organs of fish: the tuberous organs of *Eigenmannia*. J Morphol 119: 51–80
Wachtel AW, Szamier RB (1969) Special cutaneous receptor organs of fish. IV. Ampullary organs of the non-electric catfish, *Kryptopterus*. J Morphol 126: 291–308
Wagner HJ (1973) Darkness-induced reduction on the number of synaptic ribbons in fish retina. Nature (New Biology) 246: 53–55
Waltman B (1966) Electrical properties and fine structure of the ampullary canals of Lorenzini. Acta Physiol Scand Suppl 264: 1–60
Wang LDL, Boyarsky LL (1979) Release of L-glutamate from the excitatory nerve terminals of the crayfish neuromuscular junction. Life Sci 24: 1011–1014
Watkins JC (1978) Excitatory amino acids. In: McGeer EG, Olney JW, McGear PLM (eds) Excitatory amino acids. Kainic acid as a tool in neurobiology. Raven, New York, pp 37–69
Watkins JC (1981) Pharmacology of excitatory amino acid receptors. In: Roberts PJ, Storm-Mathiesen J, Johnston GAR (eds) Glutamate: transmitter in the central nervous system. Wiley, Chichester, pp 1–24
Watkins JC, Evans RH (1981)) Excitatory amino acid transmitters. Annŭ Rev Pharmacol Toxicol 21: 165–204
Watkins JC, Olverman HJ (1987) Agonists and antagonists for excitatory amino acid receptors. Trends Neurosci 10: 265–273
Watkins JC, Krogsgaard-Larsen P, Honore T (1990) Structure-activity relationships in the development of excitatory amino acid receptor agonists and competitive antagonists. Trends Pharmacol Sci 11: 25–33
Watson S, Abbot A (1990) TiPS Receptor nomenclature supplement. Elsevier Trends J 11: 1–30
Weiter CT, Nystrom B, Hamberger A (1979) Characteristics of glutamine vs glutamate transport in isolated glia and synaptosomes. J Neurochem 32: 559–565
Wenthold RJ (1985) Biochemical and immunochemical approaches to the study of the auditory system. J Acoust Soc Am 78: 334–339
Werman R (1966) Criteria for identification of a central nervous system transmitter. Comp Biochem Physiol 18: 745–766
Wersäll J (1956) Studies on the structures and innervation of the sensory epitheliane of the cristal ampullares in the guinea pig. Acta Oto-Laryngol Suppl 126: 1–85
Wersäll J, Bagger-Sjöbäck KD (1974) Morphology of the vestibular sense organs. In: Kornhuber HH (ed) Vestibular system, part I. Basic mechanisms. Springer, Berlin Heidelberg New York, pp 123–170 (Handbook of Sensory Physiology, vol VI/1)
Wersäll J, Flock Å (1964) Suppression and restoration of the microphonic output from the lateral line organ after local application of streptomycin. Life Sci 3: 1151–1155
Wersäll J, Flock Å, Lundquist P-G (1965) Structural basis for directional sensitivity in cochlear and vestibular sensory receptors. Cold Spring Harbor Symp Quant Biol 30: 115–145
Wever EG (1974) The evolution of vertebrate hearing. In: Keidel WD, Nef WD (eds) Auditory system. Springer, Berlin Heidelberg New York, pp 423–454 (Handbook of sensory physiology), vol V/1)
Wickelgren WO (1977) Physiological and anatomical characteristics of reticulospinal neurones in lamprey. J Physiol (Lond) 270: 89–114
Wickham MG (1973) Large inclusions in receptor cells of the amphibian lateral line. Am J Anat 138: 465–480
Wiet GJ, Godfrey DA, Ross CD, Dunn JD (1986) Quantitative distributions of aspartate aminotransferase and glutaminase activities in the rat cochlea. Hear Res 24: 137–150
Winberry SL, Bobbin RP (1983) Actions of acetylcholine and carbachol on the spontaneous activity of afferent fibers in the *Xenopus laevis* lateral line. 6th Meet Assoc Res Otolaryngol, Florida, St Petersburg , Beach p 47

Wissing H, Braun HA, Schäfer K (1988) Dynamic response characteristics of the ampullae of Lorenzini to thermal and electrical stimuli. In: Hamann W, Iggo A (eds) Transduction and cellular mechanisms in sensory receptors. Prog Brain Res 74: 99–107

Yamada Y, Hama L (1972) Fine structure of the lateral-line organ of the common eel, *Anguilla japonica*. Z Zellforsch 124: 454–464

Yamamura HI, Enna SJ, Kuhar MJ (1978) Neurotransmitter receptor binding. Raven, New York

Yamashita T, Amano H, Harada N, Su Z, Kumazawa T, Tsunoda Y, Tashiro Y (1990) Calcium distribution and mobilization during depolarization in single cochlear hair cells. Acta Oto-Laryngol 109: 256–262

Yamashita T, Amano H, Ohtani M, Harada N, Kumazawa T (1991) Isolation of and calcium kinetics in cochlear inne hair cells of the guinea pig. Acta Oto-Laryng 111: 879–884

Yolikoski J, Lehtosalo J (1985) Neurochemical basis of auditory fatigue: a new hypothesis. Acta Oto-Laryngol 99: 353–363

Young AB, Fagg GE (1990) Excitatory amino acid receptors in the brain membrane binding and receptor autoradiographic approaches. Trends Pharmacol Sci 11: 126–133

Zakon HH (1986) The electroreceptive periphery. In: Bullock TH, Heiligenberg W (eds) Electroreception. Wiley, New York, pp 106–156

Zhadan GG, Zhadan PH (1975) Effects of sodium, potassium and calcium ions on the functional characteristics of the catfish electroreceptors. Neurophysiology (USSR) 7: 403–410 (in Russian)

Zenner HP, Zimmermann U, Schmitt U (1985) Reversible contraction of isolated mammalian cochlear hair cells. Hear Res 18: 127–133

Zieglgänsberger W (1980) Peptides in the regulation of neuronal function. In: Bloom FE (ed) Peptides: integrations of cell and tissue function. Raven, New York, pp 219–233

Zimmerman D McG (1979) Onset of neuronal function in the lateral line. Nature 282: 82–84

Zucca G, Valli P, Casella C (1982) Ionic mechanisms sustaining activity in ampullar receptors of the frog. Acta Oto-Laryngol 93: 355–363

Zucca G, Botta L, Milesi V, Dagani F, Valli P (1992) Evidence for L-glutamate release in frog vestibular organ. Hear Res 63: 52–56

Zucca G, Vega R, Botta L, Pérez ME, Valli P, Soto E (1992) Streptomycin blocks the afferent synapse of the isolated semicircular canals of the frog. Hear Res 59: 70–74

Subject Index

MIX
Papier aus verantwortungsvollen Quellen
Paper from responsible sources
FSC® C105338

If you have any concerns about our products,
you can contact us on
ProductSafety@springernature.com

In case Publisher is established outside the EU,
the EU authorized representative is:
Springer Nature Customer Service Center GmbH
Europaplatz 3, 69115 Heidelberg, Germany

Printed by Libri Plureos GmbH
in Hamburg, Germany